6th Meeting of the Spanish Neutron Scattering Association

(SETN 2012)

Journal of Physics: Conference Series Volume 549

Segovia, Spain
24 – 27 June 2012

Editors:

J. Hernandez-Velasco
C. Cabrillo
M. Castellote
J. Ruiz-Hervias

ISBN: 978-1-63439-663-9
ISSN: 1742-6588

Printed from e-media with permission by:

Curran Associates, Inc.
57 Morehouse Lane
Red Hook, NY 12571

Some format issues inherent in the e-media version may also appear in this print version.

Printed by Curran Associates, Inc. (2015)

For permission requests, please contact the Institute of Physics
at the address below.

Institute of Physics
Dirac House, Temple Back
Bristol BS1 6BE UK

Phone: 44 1 17 929 7481
Fax: 44 1 17 920 0979

techtracking@iop.org

Additional copies of this publication are available from:

Curran Associates, Inc.
57 Morehouse Lane
Red Hook, NY 12571 USA
Phone: 845-758-0400
Fax: 845-758-2634
Email: curran@proceedings.com
Web: www.proceedings.com

TABLE OF CONTENTS

Journal of Physics: Conference Series **549** (2014) 011001 doi:10.1088/1742-6596/549/1/011001

Journal of Physics: Conference Series **549** (2014) 011001 doi:10.1088/1742-6596/549/1/011001

Preface

The bi-annual Meeting of the Spanish Neutron Scattering Association, VI RSETN, took place in the magnificent world heritage ancient city of Segovia, Spain, from 24–27 June 2012, at the historical building "Palacio de Mansilla". It was the sixth in a series of successful scientific meetings, beginning in 2002 (San Sebastián), and followed by conferences in Puerto de la Cruz (Canary Islands, 2004), Jaca (Aragón, 2006), Sant Feliú de Guixols (Cataluña, 2008) and Gijón (Asturias, 2010).

The conference covered a broad range of topics related to the use of neutron scattering techniques, from soft matter and biosciences to magnetism, condensed matter as well as advanced neutron instrumentation and applications. In addition to those topics, Spanish scientists working at neutron facilities reported recent upgrades of neutron instruments.

The VI RSETN was organized by a group of research scientists belonging to different institutions in Madrid: CSIC, Universidad Complutense and Universidad Politécnica de Madrid, in cooperation with the Spanish Society for Neutron Techniques (SETN, 'Sociedad Española de Técnicas Neutrónicas').

The meeting attracted around 90 participants. The total number of oral presentations was 36, including plenary and invited talks, both from domestic and foreign speakers. In addition, the number of posters was around 20. The success of the VI RSETN was due to the efforts of many colleagues involved at all stages of the meeting. We would like to thank the scientific committee, the local organizing committee, the chairs of the conference sessions as well as all the reviewers who agreed generously to help with the process. We would also like to emphasize the excellent scientific quality of all the presentations and posters, and we thank the support received from our sponsors (SETN, ICMM-CSIC, ESS-Bilbao, ILL, Carburos Metálicos), which was really important for the conference success.

Finally, we hope that the readers will enjoy the 28 scientific contributions contained in the present volume, which give an overview of the science and the engineering currently done by the Spanish neutron scattering community, a lively and growing community.

Jorge Hernández-Velasco, Carlos Cabrillo, Marta Castellote, Jesus Ruiz-Hervias
Conference Proceedings Editors

Scientific Committee

Regino Sáez Puche (UCM, Madrid), Marta Castellote Armero (IETcc-CSIC, Madrid), Jorge Hernández Velasco (ICMM-CSIC, Madrid), Carlos Cabrillo García (IEM-CSIC, Madrid), Jesús Ruiz Hervias (UPM, Madrid), Gaspar González Doncel (CENIM-CSIC, Madrid), Enrique López Cabarcos (UCM, Madrid), Javier Campo Ruiz (ICMA-CSIC, Zaragoza), Pedro Gorria (Univ. Oviedo), Carlos Frontera (ICMAB-CSIC, Barcelona), Miguel Angel González (ILL, Grenoble), Jon Gutiérrez Echevarría (EHU/UPV, Bilbao), Victoria García Sakai (ISIS, Chilton), Vicente Recarte (UPN, Pamplona), Antonio Urbina (UPC, Cartagena).

Organizing Committee

Regino Sáez Puche (*Chairperson*, UCM, Madrid), Marta Castellote Armero (*Secretary*, IETcc-CSIC, Madrid), Jorge Hernández Velasco (ICMM-CSIC, Madrid), Carlos Cabrillo García (IEM-CSIC, Madrid), Jesús Ruiz Hervias (UPM, Madrid), Gaspar González Doncel (CENIM-CSIC, Madrid), Enrique López Cabarcos (UCM, Madrid).

Editors of the Proceedings

Jorge Hernández Velasco, Carlos Cabrillo García, Marta Castellote, Jesús Ruiz Hervias.

Plenary speakers

Markus Strobl (European Spallation Source, Lund, Sweden), Juan Colmenero (Donostia International Physics Center, San Sebastián, Spain), Hermenegildo García (Universidad Politécnica de Valencia and Instituto de Tecnología Química, Spain), J. Paul Attfield (University of Edinburg, U.K.).

Invited speakers

José Luís García Muñoz (ICMAB-CSIC Barcelona), Gabriel Cuello (ILL Grenoble), Josep Nogués (Univ. Autónoma Barcelona), Fernando Palacio (ICMA-CSIC Zaragoza), Emilio Lorenzo (CNRS-Grenoble), Marisa Medarde (Paul Scherrer Institut), José Luís Martínez (ILL Grenoble), Félix Fernández Alonso (ISIS-STFC), Esteban Climent Pascual (Univ. Carlos III Madrid), Jesús Blanco (Univ. Oviedo), Javier Sánchez-Benítez (Univ. Complutense Madrid), Patricia Lazpita (EHU/UPV), Mónica Jiménez (ILL Grenoble), Jorge Rubio Retama (Univ. Complutense Madrid), Margarita Valero (Univ. de Salamanca), Fabienne Barroso (Centro de Física de Materiales, CSIC-EHU/UPV), Fernando Sordo (ESS-Bilbao), Federico Mompeán (ICMM-CSIC Madrid), Ricardo Fernández (CENIM-CSIC).

Journal of Physics: Conference Series **549** (2014) 011002 doi:10.1088/1742-6596/549/1/011002

Peer review statement

All papers published in this volume of *Journal of Physics: Conference Series* have been peer reviewed through processes administered by the proceedings Editors. Reviews were conducted by expert referees to the professional and scientific standards expected of a proceedings journal published by IOP Publishing.

Journal of Physics: Conference Series **549** (2014) 012001 doi:10.1088/1742-6596/549/1/012001

Baseline design of a low energy neutron source at ESS-Bilbao

F. Sordo[1,2], F. Fernandez-Alonso[3], Miguel A. Gonzalez [4], A. Ghiglino[1,2], M. Magán[1,2], S. Terrón[1,2], F. Martínez[1,2], JP. de Vicente[1,2], R. Vivanco[1,2], F.J. Bermejo[1,5], J.M. Perlado[2]

[1] Consorcio ESS-Bilbao, Paseo Landabarri 2° E-48940, Leioa, Biscay

[2] Instituto de Fusión Nuclear, José Gutierrez Abascal 2, 20886 Madrid

[3] ISIS Facility, Rutherford Appleton Laboratory, Chilton, Didcot, Oxfordshire OX11 0QX, United Kingdom

[4] Institut Laue Langevin, 6 rue Jules Horowitz 38042, Grenoble Cedex 9, France

[5] Consejo Superior de Investigaciones Cientficas, Instituto de Estructura de la Materia, Serrano 121, 28006 Madrid

E-mail: `fernando.sordo@essbilbao.org`

Abstract. This article briefly describes the basic design of the ESS-Bilbao neutron target station as well as its expected neutronic performance. The baseline engineering design, associated ancillary systems, and plant layout for the facility is now complete. A rotating target composed of twenty beryllium plates has been selected as the best choice in terms of both neutron yield and engineering complexity. It will provide neutron beams with a source term of 10^{15} $n\ s^{-1}$ resulting from the direct $^{9}Be(p,xn)$ reaction using a 75 mA proton beam at 50 MeV. The design envisages a target station equipped with two fully optimized moderators capable of withstanding a proton-beam power of 112 kW. This design is flexible enough to accommodate future upgrades in final proton energy. The envisaged neutron-beam brightness will enable several applications, including the use of cold and thermal neutrons for condensed matter research as well as fast-neutron irradiation studies. We close by discussing the role that this facility may play once the European Spallation Source becomes operational in Lund, Sweden.

1. Introduction

The design, construction, and optimization of last-generation neutron facilities such as the Spallation Neutron Source (SNS) at the Oakridge National Laboratory (ORNL, USA) and the Materials and Life Sciences Facility at the Japan-Proton-Accelerator-Complex (MLSF-JARC) has involved the activity of smaller research laboratories and/or research units within established national laboratories. The increasingly important role of these smaller institutions has been facilitated by advances in instrumentation technologies enabling the emergence of local facilities within university-scale organizations. These facilities have carried out critical work so as to render the ultimate performance of high-power sources truly competitive. The scientific and technical relevance of these smaller-scale facilities has been emphasized in an International Atomic Energy Agency (IAEA) (report in 2004 [1], highlighting the relatively unexploited synergy between high- and low-to-medium-power neutron sources. As discused in this report, the rationale behind this synergy stems primarily from the high costs associated with the operation of high-power sources and the pressing need to deliver intense neutron beams to a wide user base for as many days per year as budgetary constraints allow. These stringent requirements pose a number of difficulties for the development and testing of equipment critically needed to fully exploit their capabilities, as such facilities usually leave little room for ad-hoc tests or fully fledged research and development programmes requiring extensive access to beamtime.

The purpose of the present article is twofold. First and foremost, to describe progress to date in the design of the ESS-Bilbao facility, including its potential to host several applications in neutron science. Second and final, to identify specific areas where ESS-Bilbao will be of direct relevance for the further development of the wider European Spallation Source (ESS) project. We emphasize from the outset that the *raison d'etre* of ESS-Bilbao is entirely compatible with current ESS developments in Lund (Sweden), as current efforts at ESS-Bilbao address issues which are geared towards the effective exploitation of ESS by the Spanish neutron user community in the foreseeable future. The Spanish user base has experienced a very significant increase in size over the period 1987-2005, representing a total use of available beam time at the Institut Laue-Langevin (ILL) of approximately 5%. Notwithstanding the above, the development of this user base [2] has not been accompanied by parallel developments in expertise associated with innovative uses of the available techniques or the development of cutting-edge neutron instrumentation. A relatively brief partnership with the ISIS Pulsed Neutron and Muon Source over the period 2004-2010 provided access to ~2% of the available beamtime for the Spanish user community. Cooperation between ISIS and Spain has also represented a veritable quantum jump in the capabilities of the Spanish scientific community and industrial sector to provide state-of-the-art neutron instrumentation. In addition to the delivery of world-class neutron instrumentation including the development of scintillation-detection systems using cutting-edge technologies, this fruitful collaboration has enabled several research groups as well as local companies to develop significant capabilities in accelerator science and technology as well as spallation targets.

On the basis of the above considerations, it seems clear that participation in remotely located large-scale infrastructures such as ESS-Lund without parallel and uninterrupted developments of national capacity in instrument and technique development for subsequent deployment and exploitation at high-power sources risks mimicking the conditions of the 1987-2005 period mentioned above, ultimately leading to the suboptimal use of high-power neutron sources by Spanish scientists in the years to come. It is in this spirit that ESS-Bilbao has progressed the design and construction of a laboratory on the main campus of the University of the Basque Country near Bilbao. The main characteristics and aims of this laboratory are described below.

2. The ESS-Bilbao neutron source

2.1. The ESS-Bilbao Accelerator

The first stages of room-temperature proton and H^- accelerators are currently under construction at ESS-Bilbao. Both designs make extensive use of accelerator structures developed in close collaboration with ongoing projects such as the Accelerator Design Update at ESS-Lund, the Front End Test Stand at ISIS, and the Linac4 project at European Organization for Nuclear Research (CERN) [3]. The main accelerator parameters are listed in Table 1.

2.2. The Rotating Target Concept

Account made of the accelerator parameters listed in Table 1, a neutron target station designed so as to deliver moderated and fast neutron beams with useful fluxes for scientific applications will extend the testing capabilities and remit of ESS-Bilbao. In our case, such a target concept relies upon the neutron-yield properties of the $^9Be(p,n)$ direct reaction. The expected integrated

Journal of Physics: Conference Series **549** (2014) 012001 doi:10.1088/1742-6596/549/1/012001

Proton Energy	50 – 60 MeV
Peak Current	75 mA
Repetition Rate	20 – 50 Hz
Pulse Length - variable	0.1 – 1.5 ms
Average Current	2.25 mA

Table 1. Main parameters for the ESS-Bilbao accelerator.

neutron flux amounts to $\sim 10^{15}$ n s^{-1}, an adequate level of particle production for the testing of neutron-instrumentation components (including moderators, neutron guides, and pulse-shaping devices), as well as for the operation of several irradiation and scattering instruments. The rationale behind the development of a facility of these characteristics has been emphasized repeatedly in several IAEA official reports [4] [5]. Moreover, significant research activities and experiments are nowadays being carried out with compact sources delivering fluxes in the range $\sim 10^{13}$-10^{14} $n\ s^{-1}$ in close collaboration with large spallation facilities [6] [7] [8] [9].

The choice of a beryllium target is based on its reasonably high neutron yield of $6.5 \cdot 10^{-2}$ neutrons per proton at a beam energy of 50 MeV [10]. With the expected beam current and duty cycle, the total calculated neutron yield reaches ca. $9.4 \cdot 10^{14}$ $n\ s^{-1}$. Other options such as Li-based targets have been dismissed on the basis of significant engineering complications associated with the handling of these materials.

In the case of low-power spallation sources, targets are typically assembled by lining up a series of tungsten, tantalum, or clad uranium plates with thicknesses increasing from about a millimeter up to one centimeter or so, and with gaps of about 1 mm to allow water flow at moderate rates. Contrary to this situation, the low penetration depth of 50-MeV protons in beryllium (ca. 1.5 cm) rules out the use of such vastly proven target configuration. The use of a single plate as target has also been ruled out because of heat-load considerations. For a proton beam of diameter 10 cm, a single beryllium plate would require an average power flux in the water-cooling channel of roughly ~ 7 MW/cm^2, a figure that is far beyond the limits of the state-of-the-art for low-pressure water-cooling systems. We note that this value is roughly twice the heat flux estimated for the ESS-2003 Solid Target Project Proposal [12].

The solution proposed by ESS-Bilbao to circumvent these difficulties capitalizes on previous developments in high-power spallation targets. The use of a rotating target wheel with 20 beryllium plates ensures that each plate sees the proton beam at an effective repetition frequency of 1 Hz. In this configuration, the total heat load on each element is limited to 5.6 kW, and the time-averaged heat flux is 0.4 kW/cm^2. Although the average heat flux may not be a problem, the instantaneous heat flux during the pulse can be considerably higher, and thus the maximum temperature in the water-cooled surface can quickly approach 100 ^{o}C. A water pressure of 5 bar is sufficient to withstand this temperature rise. Using the value for the water pressure just referred to, simulations carried out using the FLUENT-ANSYS packages predict a maximum stress in the range of 80 MPa, a value well below the fatigue limits of beryllium. Figure 1 shows the average temperatures and von Mises stresses on the target.

Figure 2 shows the configuration of the target-moderator-reflector (TMR) assembly. The beryllium plates are embedded in a wheel rotating on the horizontal axis and in contact with the water channels. The proton beam hits the target at a relative angle of incidence with respect to the plates of 45°. On the other side of the target assembly, a methane moderator is placed in order to maximize neutronic performance. Both target and moderator are surrounded by the reflector. Around the reflector, several layers of shielding material are used to minimize dose rates outside the target station.

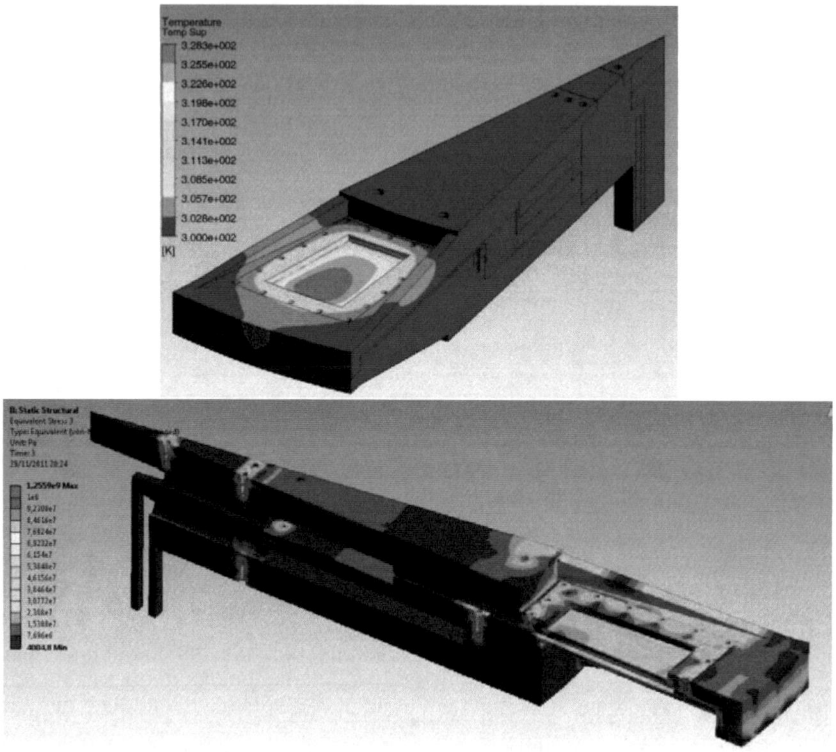

Figure 1. Three-dimensional representation of average temperatures and von Mises stresses in the target assembly.

2.3. Neutronic Performance of the Target Station

Since the ESS-Bilbao source provides moderate neutron intensities, detailed optimization of the moderator-re ector assembly is a must. There are three main con gurations for positioning the moderators with respect to the neutron-producing target, namely: WING, SLAB, and Flux TRAP [11]. In general, a WING con guration results in a lower brightness than a SLAB con guration, but since moderators are usually located above and below the target element, this particular geometry precludes the use of neutron-transport devices such as inserts and beam tubes in line of sight with the target through the moderator. The latter limitation is of concern for the SLAB con guration where moderators are located sideways to the target and thus may lead to signi cant leakage of fast neutrons coming from the direct view of the target. These considerations are of particular relevance for high-power spallation sources since the relevant nuclear reactions display a rst step or hadron cascade leading to a highly anisotropic emission of high energy neutrons, while a second step involving ssion/evaporation leads to the isotropic emission of high-energy neutrons[13]. For this reason, the majority of high-power sources like ESS [14], ISIS [15], or SNS [16] prefer the use of WING or Flux-TRAP con gurations to reduce backgrounds associated with fast-neutron production.

Nevertheless, for $^{9}Be(p\ n)$ reaction at 50 MeV, most of the high energy neutrons are produced in the forward direction [17]. Hence, background neutrons can be controlled by creating a 45° angle between beam and experimental lines. Considering this factor, the slab con guration will maximize neutron production.

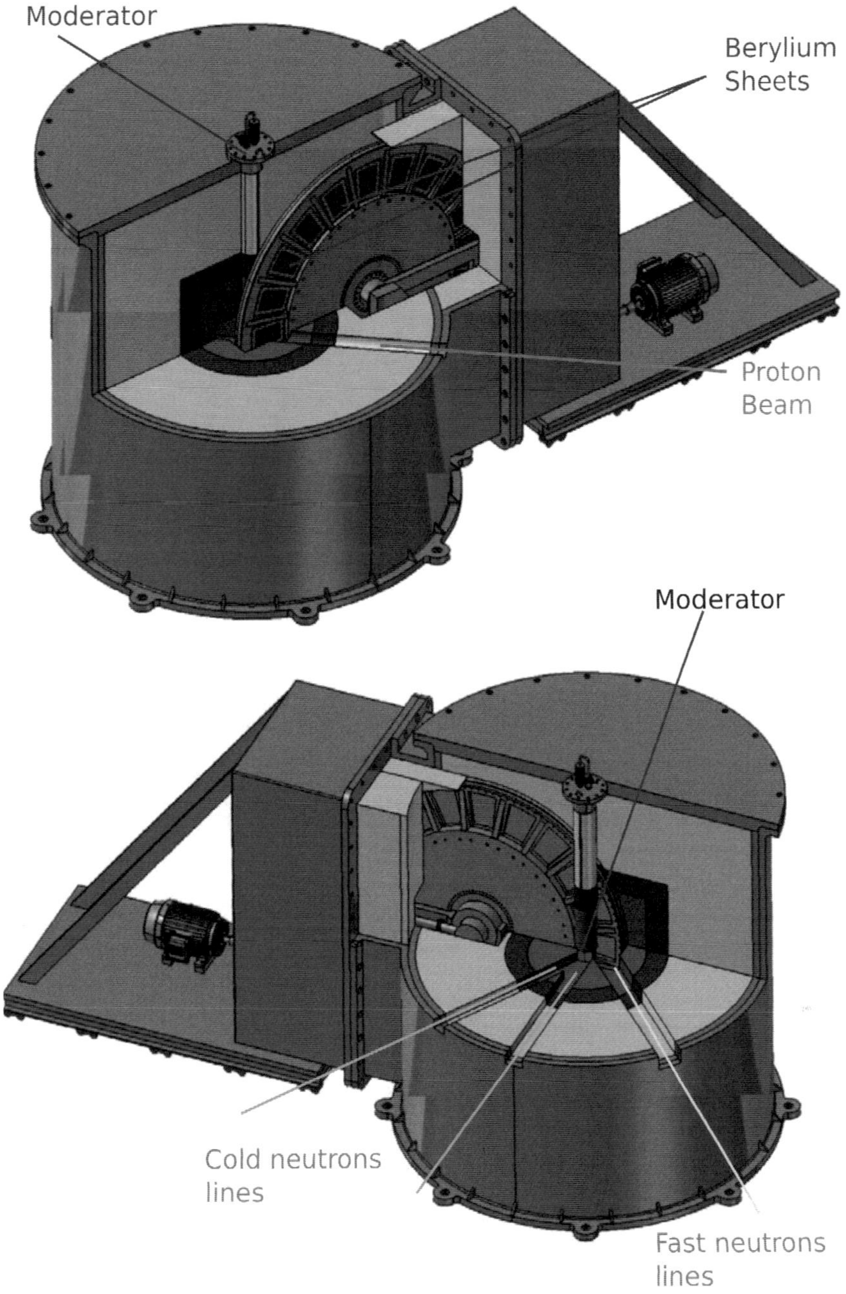

Figure 2. ESS-Bilbao Target-Moderator-Re ector assembly. For further details see the text.

Figure 3 shows the proposed TMR layout with a solid methane moderator in SLAB geometry.

Methane has been selected as moderator material because of its excellent neutronic performance [7] as well as the extensive experience gained from years of operation at existing neutron sources. The long-pulse nature of the installation (maximum pulse length of 1.5 ms) dictates that a coupled moderator represents the best choice [8] to maximise neutron production. The methane moderator and target are also surrounded by a re ector material to increase neutron brightness on the moderator surface. Current estimates of heat load on the moderator under operating conditions are ca. 50 W. Although signi cantly smaller than the value of 300 W for the ISIS-TS2 coupled moderator [18], it is still not a negligible gure. Some radiation chemistry e ects such as radiolysis and polymerization processes are expected to occur, although previous experience has paved the way to solve these problems by means of periodic annealing and thermal cycling protocols. Other options currently under consideration for the second moderator include the use of liquid hydrogen or mesitylene.

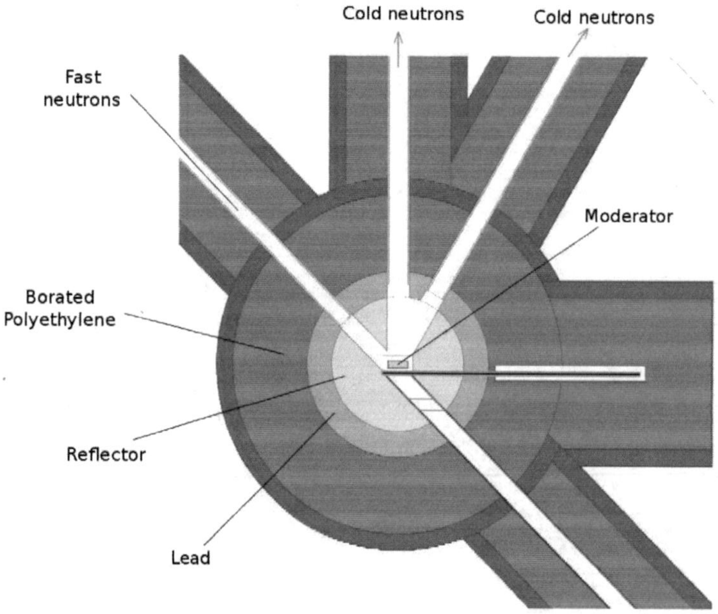

Figure 3. ESS MCNPX model of the TMR assembly with one moderator in SLAB geometry.

Values for the main TMR parameters were obtained via an optimization process to maximize neutronic performance. The Figure 3 shows the MCNPX model develop for the process. Optimal values include a 12 10 2.5cm^3 methane moderator, a 40 cm thick beryllium re ector, and a distance of 8 cm from the target surface to the moderator center (mainly needed because of target engineering constraints). The re ector is surrounded by several layers of shielding materials: 15 cm lead, 60 cm of borated polyethylene, and 10 cm of steel. The rst layer of lead aims to shield against gamma radiation close to the center while minimizing the total weight of the shielding. The borated polyethylene will moderate and absorb most neutrons. Finally, the steel shielding will absorb gamma radiation resulting from neutron capture in the borated polyethylene. This con guration aims to reduce the activation of the external elements as much as possible with the least amount of shielding.

Several possibilities have been explored for the re ector material and a summary of the results is shown in Figure 4. This figure shows the distribution at the moderator surface of ~ 5 meV neutrons for long proton pulses (1.5 ms) and different reflector materials. The beryllium reflector introduces significant moderation, has a very low threshold for (n, xn) reactions, and does not have a large capture cross section. The result of this combination is a maximum neutron production associated with a very long tail in the time distribution. Lead has a very high density, not a very high threshold for (n, xn) reactions, and has very poor moderating properties. The combination of these factors results in a much lower neutron yield and a short tail in the time distribution. Alternatively, heavy water provides low neutron absorption and significant scattering cross section but, as a drawback, it has not (n, xn) reactions. Its ultimate performance is therefore low. Finally, a light-water reflector displays a very high scattering and absorption cross sections. As a result, its performance sits right in between lead and beryllium.

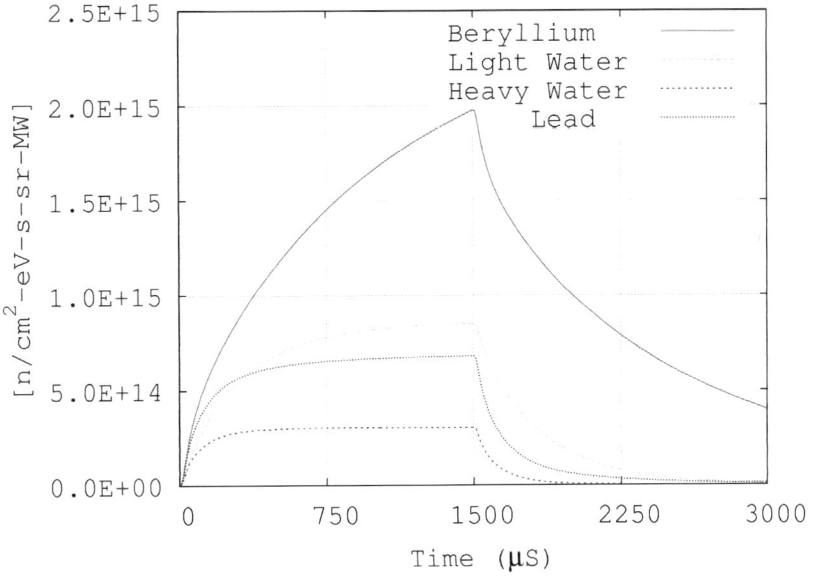

Figure 4. Long pulse for ~ 5 meV neutrons (integration range: 4.4-5.3 meV).

The water reflector exhibits some interesting properties which warrant further discussion. Figure 5 shows the neutron flux on the moderator surface arising from the water reflector compared to a beryllium reflector scaled by a factor of 0.325. These two results are also compared to the predicted ESS time distribution [19] for a 1.5 ms proton pulse. It is rather remarkable that the water reflector on ESS-Bilbao is nearly identical in shape to that predicted for ESS. The scaling factor between these two sources amounts to ~ 800.

The SLAB configuration for a solid methane moderator and beryllium reflector shows very good neutronic performance. The source term ratio between ESS and ESS-Bilbao differs by a factor of ~ 750. The average ratio in neutron brightness is close to a factor of $\sim$160/100 lower without/with beryllium filter. The proposed ESS-Bilbao target station is therefore considerably more efficient per source neutron than ESS.

As reference, this medium-size neutron source will be 20 times more intense than Low Energy Neutron Source (LENS) [20] and ca. 5 times below similar moderators at ISIS-TS2. Table 2.3 shows a comparison between ESS-Bilbao parameters and data from other international facilities [21].

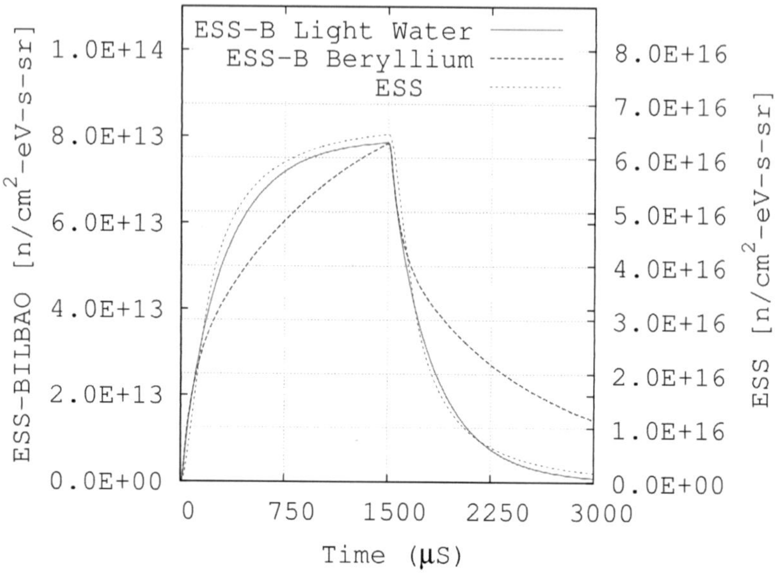

Figure 5. Time distribution of ~ 5 meV neutrons for a long proton pulse (integration range: 4.4-5.3 meV). The beryllium brightness has been scaled by a factor 0.325. For further details, see the text.

	Power	Rep Rate Hz	Viewed Surface $[cm^2]$	N. intensity $[n/cm^2 \cdot s \cdot sr]$	N. intensity $[n/cm^2 \cdot pulse \cdot sr]$
JSNS C. hydrogen	300	25	10 w x 10 h	1,3E+12	5,1E+10
	1000	25	10 w x 10 h	4,5E+12	1,8E+11
SNS C. hydrogen	1000	60	12 w x 10 h	2,1E+12	3,5E+10
	1400	60	12 w x 10 h	3,0E+12	4,9E+10
ISIS-TS2 H/CH4, gro.	48	10		5,0E+11	5,4E+10
ISIS-TS2 H/CH4, hyd.	48	10	12 w x 11 h	3,0E+11	3,0E+10
ESS-B	112	20	12 w 10 h	1,3E+11	6,6E+09

Table 2. ESS-Bilbao neutron source compared to other international facilities

With the above figures in mind, we conclude that a mid-intensity neutron source at ESS-Bilbao will enable a sufficient flux of cold and thermal neutrons for component testing, the development of new instrumentation, and the execution of neutron-scattering experiments in condensed matter. At the time of writing, such capabilities are to be found nowhere in Spain.

3. Neutron instrumentation: preliminary considerations

The main parameters for the ESS-Bilbao accelerator given in Table 1 provide lower and upper bounds for the temporal pulse length of the proton beam, where short(long)-pulse operation corrresponds to a pulse width of 0.1(1.5) ms and a repetition rate of 50(20) Hz. In what follows, we consider these two limits of operation for the effective use of cold and thermal neutrons in condensed matter research. To provide a realistic assessment of future capabilities at ESS-Bilbao, we shall limit the present discussion to well-established and thoroughly tested instrument concepts at pulsed (mostly spallation) neutron sources over the past three decades.

Following Windsor [22], we also make a clear distinction between *total* vs. *useful* neutron flux. The efficient use of a pulsed neutron source involves the judicious exploitation of its inherent time structure, enabling the use of a sufficiently wide dynamic range of incident wavelengths (multiplexing advantage) with the requisite instrumental resolution (R). In this situation, R is primarily dictated by the ability to discriminate between neighbouring incident neutron wavelengths in the time domain, that is, $\Delta\lambda/\lambda \propto \Delta t/L$. Δt is the temporal pulse width of the moderated neutron (a wavelength-dependent quantity) and L is the distance between source and sample. Optimal instrument design seeks to match incident and scattered components to the resolution function in order to maximise final detected flux for a given R [22].

In the same spirit as Schober et al. [23], it is convenient to consider several neutron-instrument classes in terms of their resolution characteristics. These categories are briefly described below, along with a summary of specific (yet by no means exhaustive) areas of scientific application:

- Coarse (R ~1): irradiation studies, activation analysis, boron capture therapy, basic neutronography, neutronics (component testing, moderator studies).
- Low (R ~10^{-1}): low-angle scattering, reflectometry, spin-echo techniques.
- Medium (R ~10^{-2}): diffraction, energy-resolved imaging, broadband spectroscopy.
- High (R ~10^{-3}): high-resolution (e.g., back-scattering) spectroscopy and diffraction.

Figure 6 provides a summary of instrumental characteristics in terms of dynamic range and resolution for ESS-Bilbao as a long- and short-pulse neutron source. We assume thermal and cold neutrons have been produced by the cold methane moderator described in earlier Sections of this paper. Scientific applications requiring coarse (R ~1) resolution are equally possible in both cases, as all of these do not rely on the use of the temporal structure of the neutron pulse and, if required, energy selection may be performed by some other means (e.g., monochromator, choppers).

As a long-pulse source (1.5 ms at 20 Hz), the data shown in Figure 6 shows that the optimal compromise between resolution and dynamic range requires instrument lengths of ~ 30-40 m. At these distances, the resulting resolution amounts to R ~10^{-1}, typical for low-angle and neutron reflectivity instrumentation. To attain medium resolution (R ~10^{-2}), distances in excess of 100 m are required. These figures are to be contrasted with a maximum attainable instrument length of ~50-60 m imposed by space constraints in the current facility layout for ESS-Bilbao. With these limitations in mind, we conclude that 1.5 ms pulses will enable the exploitation of both coarse and low-resolution instrumentation using well established instrument-design concepts. A significant increase in resolution is also possible via the use of neutron spin-echo techniques [24, 25].

ESS-Bilbao operating as a short-pulse source (0.1 ms at 50 Hz) constitutes a more favorable situation (cf. Fig. 6). Efficient use of the methane moderator spectrum can be achieved at distances of ~15 m with medium (R ~10^{-2}) resolution for cold neutrons. Medium-to-high resolution is also achievable for distances of 40-50 m using well-established instrument concepts. All in all, the use of proton pulses of ~0.1 ms represents an optimal combination of dynamic range and resolution for a large proportion of instrumental applications. These pulse widths are also comparable with those typically used in high-resolution spectrometers and diffractometers at short-pulse spallation sources (e.g., the IRIS spectrometer at ISIS [26]), as well as in last-generation coupled moderators [27], an important area for further R&D efforts. Further and significant gains in resolution to reach and cross the R ~10^{-3} 'barrier' are also possible via the exploitation of pulsed-shaping choppers (PSC's) placed in the vicinity of the source. This concept has been recognized as the way forward for current and future developments in high-resolution neutron spectroscopy at JPARC [28] and ISIS [29, 30]. The new Biomolecular Dynamics

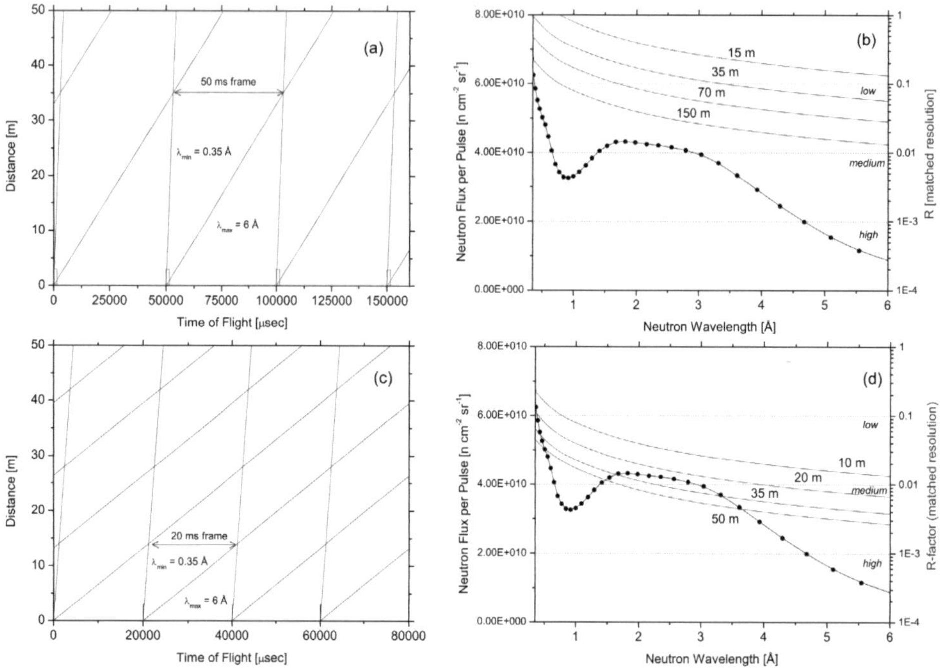

Figure 6. Time-distance diagrams (left) and resolution (right) for long- (panels a,b) vs. short-pulse operation (c,d) at ESS-Bilbao. For reference, the neutron spectrum of the cold methane moderator is also shown in (b) and (d). For further details, see the text.

Spectrometer DNA at J-PARC [31] relies entirely on the use of a PSC in order to reach a resolution of a few μeV with a relatively modest incident path of 42 m. We anticipate that ESS-Bilbao will capitalize extensively from the exploitation of these innovative technologies to tailor flux and resolution in order to meet increasingly demanding scientific requirements.

4. ESS-Bilbao as an experimental support station for ESS.

Finally, we provide a brief outline of the activities envisaged for the ESS-Bilbao site,and highlight their relevance for the full development and exploitation of ESS-Lund. At present, ESS-Bilbao profits from the experience gained through fruitful collaborations with the highest-power spallation sources in operation, namely, SNS at Oak Ridge , as well as with ISIS in the UK and with the reactor source at the Institut Laue Langevin. The rationale behind such efforts stems from the unequivocal realization that further improvements in the efficient use of neutrons is far more cost effective than net increases in neutron production. Therefore, significant efforts should be directed towards improvements in source performance to achieve (quite realistically) order-of-magnitude gains in *useful* neutron flux.

Such previous experience together with the successful symbiosis between universtity university-based sources serve to envisage clear synergies o issues such as

- *Critical work on components leading to the efficient use of spallation neutrons*, alongside the lines highlighted in recent publications [1, 32] as well as in a number of contributions to meetings of the Union for Compact Accelerator Driven Neutron Sources [33]. UCANS is an alliance between ten neutron laboratories from the Americas, Asia, and Europe. Particular areas where such synergies have already been demonstrated include the development of very cold moderator materials, modelling and measurement of neutronic properties of new cryogenic moderators, R&D of efficient pelletized cold neutron moderators, the development of novel concepts such as directional moderators or, finally, the exploration of new routes for the production of ultracold neutrons by means of novel methods involving solid oxygen or deuterium.
- *Development of new instrumentation for high-power sources.* As a matter of fact, small-to-medium power sources have played an important role in the development and testing of novel neutron instrument concepts and components. New instrumentation enabling new science is typically carried out on test beamlines at lower-power facilities in order not to compromise the tight experimental programmes at high-power sources. In this context, several advantages characteristic of CADNS can be exploited for this purpose, including the reduced shielding requirements owing to lower proton and neutron energies. This feature enables the setting-up of instrument-development stations easily adaptable for testing, development, and training. Contrary to naïve expectation, the inherent time structure of a pulsed source makes CADNS's more suitable for the development of such kind of instrumentation as opposed to continuous (mostly reactor-based) or quasi-continuous sources.
- *Use of ESS-Bilbao for the design and preparation of ESS experiments,* for users aiming to carry out feasibility tests on measurements to be proposed to be carried at ESS. The source will also serve to perform some experiments which do not necessitate the ultra-high fluxes, including those not requiring very high intensities, such as for example a significant fraction of powder-diffraction measurements aiming at materials characterization, or other innovative experiments requiring a feasibility study prior to beam-time allocation on highly oversubscribed instruments.
- *Training new users in neutron-scattering techniques.* There is an inherent difficulty in achieving this goal at large facilities because of the pressure to make the most efficient use of beamtime.

In this respect, it is worth recalling some of the conclusions from the last survey carried out by the European Neutron Scattering Association (ENSA). Specifically, Conclusion 4 states that [34]:

Neutron experiments are planned at appropriate sources. Top-tier sources are extremely important for "demanding experiments, while a large portion of "ordinary research projects can be done at medium flux or sometimes even at low flux sources with an appropriate instrument suite. Our vision is thus fully consistent with such a three-tier structure for neutron science in Europe where, in addition to the top-tier ESS and large sources such as ILL and ISIS, a number of compact national infrastructures should serve to support both the user community and future source developments. The adequacy and timeliness of such a view has already been demonstrated by the ongoing collaboration between university-based facilities and high-power sources in the USA and Japan.

References

[1] IAEA-TECDOC-1439 2005
[2] Available at http://neutron.neutron-eu.net/FILES/ENSA_2005.pdf
[3] Kronberger M, Kuchler D, Lettry J, Midttun O, Oneil M, Paoluzzi M, and Scrivens R 2010 *Review of Scientific Instruments* **81** 02A708

[4] IAEA-TECDOC-1439 2005
[5] IAEA-CRP-F12021 2012
[6] Kiyanagi Y, Watanabe N, and Iwasa H 1992 *Nuclear Instruments and Methods in Physics Research* A **312** 561-570
[7] Kiyanagi Y, Watanabe N, and Nakajima M 1994 *Nuclear Instruments and Methods in Physics Research* A **343** 550-557
[8] Kiyanagi Y 2006 *Nuclear Instruments and Methods in Physics Research* A **562** 561- 564
[9] Baxter D V, Ansell S, Ferguson P D, Gallmeier F X, Iverson E B, Kaiser H, Lu W, Muhrer G, Rinckel T C and Steinbach T 2012 *Physics Procedia* **26** 153-160
[10] Tilquin I, Froment P, Cogneau M, Delbar Th, Vervier J, and Ryckewaert G 2005 *Nuclear Instruments and Methods in Physics Research* A **545** 339-343
[11] Watanabe N 2003 *Reports on Progress in Physics* **66** 339
[12] . Bauer G S, 2010 *Journal of Nuclear Materials,* **398** 19-27
[13] Filges D and Goldenbaum F 2010 *Handbook of spallation research: theory, experiments and applications* (Weinheim: Wiley-VCH)
[14] Bauer GS 1997 *Physica* B **234236** 1214-1219
[15] Bennington S M, Campbell S I, Broome T A, Picton D J, and Beynon T D 2003 *Journal of Neutron Research* **1-2** 93-98
[16] Charlton L A, Barnes J M, Gabriel T A, and Johnson J O 1998 *Nuclear Instruments and Methods in Physics Research* A **411** 494-502
[17] Kamada S, Toshiro I, Yasuhiro U, Wataru T, Takuji O, and Mamoru B. 2011 *Journal of the Korean Physical Society* **59** 1676-1680
[18] Crispel S, Bonneton M, Thiering R, Simon M, and Teah J 2010 *Proccedings of IPACÍ0, kyoto* 412-414
[19] Magán M, Terrón S, Thomsen K, Sordo F, Perlado J M, and Bermejo F J 2012 *Nuclear Instruments and Methods in Physics Research* A **680** 61-68
[20] Lavelle C M, Baxter D V, Bogdanov A, Derenchuk V P, Kaiser H, Leuschner M B and Lone M A 2008 *Nuclear Instruments and Methods in Physics Research* A **587** 324-341
[21] Maekawa F, Harada M, Oikawa K, Teshigawara M, Kai T, Meigo S, Ooi M, Sakamoto S, Takada H, Futakawa M, Kato T, Ikeda Y, Watanabe N, Kamiyama T, Torii S, Kajimoto R, and Nakamura M 2010 *Nuclear Instruments and Methods in Physics Research* A **620** 159-165
[22] Windsor C G, 1981 *Pulsed Neutron Scattering* (London: Taylor & Francis)
[23] Schober H, Farhi E, Mezei F, Allenspach P, Andersen K, Bentley P M, Christiansen P, Cubitt B, Heenan R K, Kulda J, Langan P, Lefmann K, Lieutenant K, Monkenbusch M, Willendrup P, Šaroun J, Tindemans P, and Zsigmond G 2008 *Nuclear Instruments and Methods in Physics Research* A **589** 34
[24] Mezei F 1972 *Z. Physik* **255** 147
[25] Mezei F (ed.) 1980 *Neutron Spin Echo* (Heidelberg: Springer)
[26] Carlile C J and Adams M A 1992 *Physica* B **182** 431
[27] Watanabe N 2003 *Rep. Prog. Phys.* **66** 339
[28] Takahashi N, Shibata K, Sato T J, Tamura I, Kajimoto R, Harjo S, Oikawa K, Arai M and Mezei F 2007 *J. Phys. Chem. Solids* **68** 2199
[29] Demmel F and Andersen K H 2008 *Meas. Sci. Technol.* **19** 034021
[30] Demmel F, Garcia S V, Telling M T F, and Fernandez-Alonso F 2010 *Rutherford Appleton Laboratory Technical Reports* RAL-TR-2010-005
[31] URL: http://j-parc.jp/MatLife/en/instrumentation/bl02/img/DNA.jpg
[32] D V Baxter *et al* 2009 *Working paper on: Improved production and utilization of short pulsed, cold neutrons at low-medium energy spallation neutron sources Report of the 2nd Research Co-ordination Meeting* (Vienna: IAEA)
[33] See http://www.indiana.edu/lens/UCANS/
[34] See page 5 of Ref.[2]

Journal of Physics: Conference Series **549** (2014) 012002 doi:10.1088/1742-6596/549/1/012002

Neutron scattering at high temperature and levitation techniques

G J Cuello[1], V Cristiglio[1], L Hennet[2], I Puente-Orench[1,3]

[1] Institut Laue Langevin, 6 rue Jules Horowitz, 38042 Grenoble, France
[2] CEMHTI and University of Orléans, 45071 Orléans, France
[3] Instituto de Ciencia de Materiales de Aragón, c/ Pedro Cerbuna, 50009 Zaragoza, Spain

E-mail: cuello@ill.eu

Abstract. Studies of the liquid state present an obvious fundamental interest and are also important for technological applications since the molten state is an essential stage in various industrial processes (e.g. glass making, single crystal growing, iron and steel making).

Most of the physical properties of a high-temperature liquid are related to its atomic structure. Thus it is important to develop devices to probe the local environment of the atoms in the sample. At very high temperature, it is difficult to use conventional furnaces, which present several problems. In particular, physical contact with the container can contaminate the sample and/or modify its structural properties. Such problems encouraged the development of containerless techniques, which are powerful tools to study high-temperature melts. By eliminating completely any contact between sample and container, it is possible to study the sample with a very high degree of control and to access very high temperatures. An additional advantage of levitation methods is that it is possible to supercool hot liquids down to several hundred of degrees below their equilibrium freezing point, since heterogeneous nucleation processes are suppressed.

1. Introduction

While studies of liquids at high temperatures present an obvious interest from a fundamental point of view, they also present significant technological interest since the molten state is an essential stage in various industrial processes (glass making, semiconductor technologies, aerospace companies, iron and steel industry, etc.). The high-temperature properties of oxide minerals or iron-based liquid alloys also play a crucial role in the geophysical behavior of the Earth.

Despite a considerable number of studies performed over the past 50 years, the liquid state is still not well understood. It is therefore important to develop experimental techniques as well as numerical simulation capabilities for studying the properties of high-temperature liquids.

Studies of liquid materials at high-temperature have to contend with various problems when they are carried out in a traditional furnace where the sample is held in a container:

- Chemical reactions between the sample and the container.
- Contamination of the sample from the container.
- Difficulty of reaching temperatures above 1500°C.
- Difficulty of undercooling the sample due to heterogeneous nucleation.

The use of containerless methods avoids these problems. Over the past few years an increasing number of studies devoted to high-temperature molten materials have been carried out as a result of the development of levitation techniques. Several laboratories around the world have developed various kinds of levitation apparatus, including acoustic, electromagnetic, electrostatic and aerodynamic levitation techniques. They have been applied to determine physical properties of levitated liquids. A recent review can be found in Ref. [1]. Some of them have been combined with x-ray and neutron scattering techniques to study the structure and the dynamics of molten materials.

The CEMHTI in Orléans has chosen to combine aerodynamic levitation with a CO_2 laser heating [2, 3, 4]. The advantages of the technique are the simplicity and compactness of the device that allows its integration into different kinds of experiments. This technique makes it possible to access very high liquid temperatures while maintaining the sample's purity, and to investigate metastable states in the undercooled temperature range, down to a few hundred degrees below the equilibrium melting temperature.

2. Levitation techniques

The principle of levitation is to create an opposing force to counteract the gravity of the sample and lift it away from support structures. In the past, five main levitation methods have been combined with x-ray and neutron scattering techniques at synchrotron and neutron sources: single-axis acoustic levitation (SAL), electromagnetic levitation (EML), electrostatic levitation (ESL), conical nozzle levitation (CNL) and gas film levitation (GFL).

2.1. Electromagnetic levitation

Electromagnetic levitation is mainly suitable for electrically conductive materials. It will be discussed here only briefly since a detailed description of the EML can be found elsewhere [5, 6].

With EML a radio-frequency electromagnetic field is generated by a levitation coil and Foucault currents are induced in the sample. This leads to an inductive heating of the sample, and at the same time, the interaction of the Foucault currents with the magnetic field of the coil leads to a force that counteracts gravity and makes it possible to levitate the sample. The electromagnetic force applied on the levitated sample depends on the power absorbed by the latter. It is essentially a function of the square of the magnetic field strength and the electrical conductivity of the sample material. So by varying the heating power, one can control the temperature of the sample. The levitation coil is situated in a vacuum chamber which is first evacuated and then filled with a very high purity gas, usually He or a mixture He plus a few percent of H_2.

2.2. Electrostatic levitation

With ESL [7, 8], the sample is electrically charged and levitated in a vertical electro-static field between two electrodes. Two pairs of smaller side electrodes are used to position the sample horizontally. When it levitates, the sample is melted using lasers. With this method, it is possible to study all samples that can be electrically charged. This method has various advantages. First, it works under vacuum, preventing contamination, and allows studying poor electrical conductors or materials with a low melting point. One drawback is the complexity of the setup that limits its combination with various spectroscopies. While some insulating oxides can be studied with this technique, it is used mostly with metals.

2.3. Acoustic levitation

Acoustic levitation can suspend small objects via the acoustic radiation force that results from an impedance difference between the suspension medium, normally a gas, and a solid or liquid

Figure 1. Aerodynamics levitator for neutron di raction measurements of liquids and supercooled samples. Installation at D4 instrument at ILL.

sample. A single-axis acoustic levitator consists mainly of two parts: a vibrating source (or transducer) generating ultrasound (a few tens of kHz) and a re ector with a concave surface to improve the e ciency. The vibrating source-re ector axis is parallel to the direction of gravity and an acoustic wave is generated along this direction to counteract gravity.

SAL can be used to study low melting point alloys and non-conducting materials as well as room temperature liquids. In the past, Trinth developed a device enabling to work up to 700 C [9]. Resistive heating coils were used to heat the sample. But even though, the accessible temperature range is limited and this technique is not for very high temperature studies.

2.4. Gas lm levitation

Developped by Granier and Potard [10], gas ølm levitation enables the levitation of an object against gravitational force by oating on a thin gas ølm formed by gas ow through a porous membrane. The gas ølm thickness is typically several tens of micrometers. With this technique, it is possible to levitate a large amount of material (a few grams), which can be heated using a furnace, for instance. This technique could be also employed for viscosity measurements [11]. This is the main drawback of this levitation technique for its use in neutron scattering experiments, because the beam touches the membrane and its support. Moreover, the shape of the levitated sample is almost at, which is not well suited for di raction experiments, although it could be useful for re ectometry. The proximity of the sample to the membrane prevents the use of high temperature with GFL.

2.5. Conical nozzle levitation

Aerodynamic levitation was ørst developed in the seventies [12] then later adopted at the CEMHTI [13] for RMN studies. It has proved to be a powerful and versatile technique for studying highly reactive liquids in the equilibrium melt and the supercooled liquid state several hundred Kelvin below the melting point [14]. In particular, it is now widely used with neutron and x-ray di raction techniques to investigate the structural properties of liquids [14, 15, 16]. The working principle of aerodynamic levitation is well known and is described in detail elsewhere [17]. The basic idea is to circulate levitation gas (usually argon) through a nozzle onto the sample from below in order to counteract gravity and lift it above the nozzle. The sample is then heated to the desired temperature by means of one or two focused CO_2 lasers. The levitation gas ow is accurately regulated and monitored by a mass ow controller enabling to maintain the sample su ciently stable for long counting times.

In comparison to the other levitation techniques (e.g. applying electrostatic [7], electromagnetic [18] or acoustic [19] øelds), it has the outstanding advantage of supporting and melting any type of material ranging from insulators through semiconductors to metals. A current drawback is that samples need to be relatively small (3 mm) in order to assure aerodynamic stability, but progress is now being made in this area.

Figure 2. Aerodynamics levitator for small angle neutron scattering. A photo from measurements on D22 instrument at ILL.

3. Levitation at ILL

At the Institute Laue-Langevin (ILL) almost all previously mentioned techniques have been used on di erent instruments. The majority of such experiments have been performed in collaboration with external groups bringing the equipment to ILL. In the case of SAL and CNL techniques, the ILL sta is developing standard systems that in a near future should be available as standard environment for several instruments.

3.1. Aerodynamic levitator

An aerodynamic levitation setup has been specially designed for neutron diffraction experiments. Figure 1 is a schematic view of the levitation setup integrated into the D4c diffractometer at ILL. This device is described in detail by Hennet et al. [14] and we give here a short description of the working principle. The spherical sample is levitated in a convergent/divergent nozzle under an argon/oxygen (3%) gas flow and heated to the desired temperatures by two 125 W CO_2 lasers directed from above. The laser beams are focused on the sample by means of spherical mirrors at two different angles in order to obtain a homogeneous temperature distribution. Two NaCl windows are used to transmit the beam into the vacuum chamber. The pyrometer is placed inside the chamber to avoid window corrections. A third laser directed at the sample from below through the nozzle is used to compensate the cooling of the sample by the gas flow. A high-quality video image of the sample taken from above is continuously displayed in order to monitor the levitation of the sample during heating. Video images of the sample are also displayed with a horizontal camera in order to determine the sample position in the levitator and to monitor the vertical stability.

Figure 3. The single-axis acoustic levitator present at ILL.

For small angle neutron scattering measurements the laser-heated containerless aerodynamic levitation technique was adapted to D16 and D22 instruments at ILL. A picture of the set-up on D22 instrument for SANS measurements is showed on Figure 2. It uses two lasers, one from above and another from below, to heat a small sample (2.5 mm to 3 mm diameter) that is aerodynamically levitated about 1 mm above a small nozzle. For metallic samples, we use Ar gas whose oxygen content has been reduced to about 0.1 ppm by our gas purifier, whereas for oxide samples we use Ar with 3% oxygen content. Our setup at D22 uses stainless-steel gas-flow tubing. One pump-down and then filling with pure Ar in order to run metallic samples at high temperature for several hours without oxidation problems.

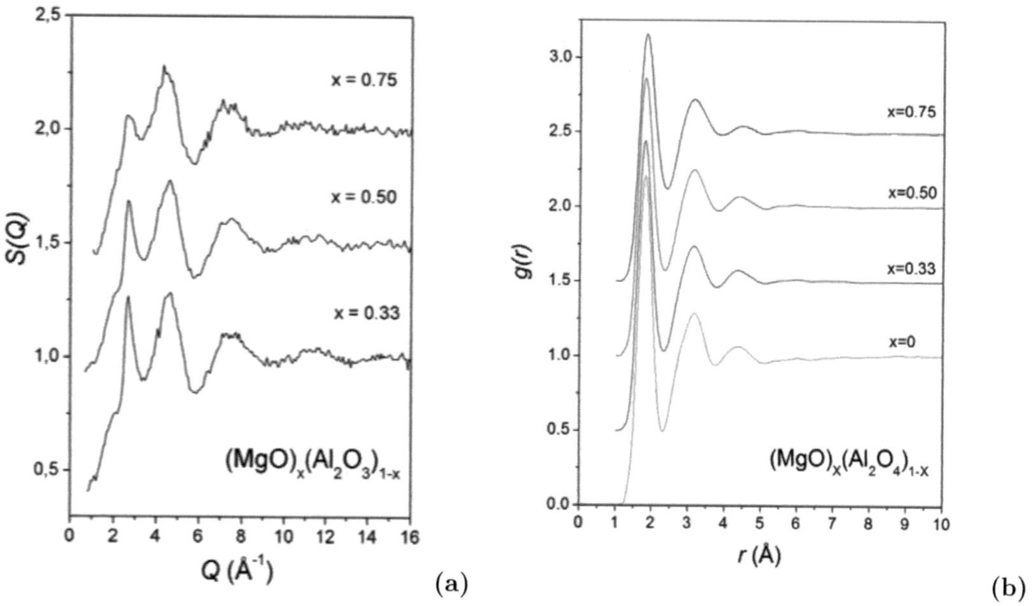

Figure 4. (a) Average structure factor $S(Q)$ for liquid $(MgO)x(Al_2O_3)_{1-x}$ at 2400 K measured on D4 at ILL. Curves have been shifted up for clarity. (b) Corresponding pair distribution functions $g(r)$. [20]

3.2. Acoustic levitator

A single-axis acoustic levitator installation has been recently used for neutron scattering measurements at ILL [19]. This apparatus consists of two vertically-opposed, high-output acoustic transducers that operate at a frequency of 22 kHz and can produce sound pressure levels to 160 dB. The produced acoustic waves are monitored from the amplifier using an oscilloscope. The levitator is composed of three main parts (Fig. 3): the levitator assembly, the ultrasonic power amplifier, and the acoustic controller. The transducers are mounted in aluminum tubes and located on a metal stand. Acoustic absorbing foam disks approximately 50 mm in diameter were glued onto the face of the transducer horns to reduce unwanted reflections that can cause instabilities in the levitated sample. Samples are introduced into the "sound field" using a small syringe or a wire gauze spoon and can be translated by adjusting the phase between the transducers or squeezed by modulating the acoustic levels with variable frequencies. To remove the sample we have simply to use the spoon for the solid samples or a flat surface for liquids. In order to avoid missing samples in the basement of the levitator a plastic film is positioned in a way to cover the entire surface. At present the levitator can only be used at room temperature. For doing experiments at low temperatures, the use of a cryostream is necessary, but this possibility is still under study. The cryostream nozzle would be arranged at approximately 50 mm to 100 mm from the sample in a way that the cooling gas is flowed over the levitated sample. The SAL is easily to install and can be integrated in several instruments present at ILL for the in-situ investigation of materials and aqueous solutions.

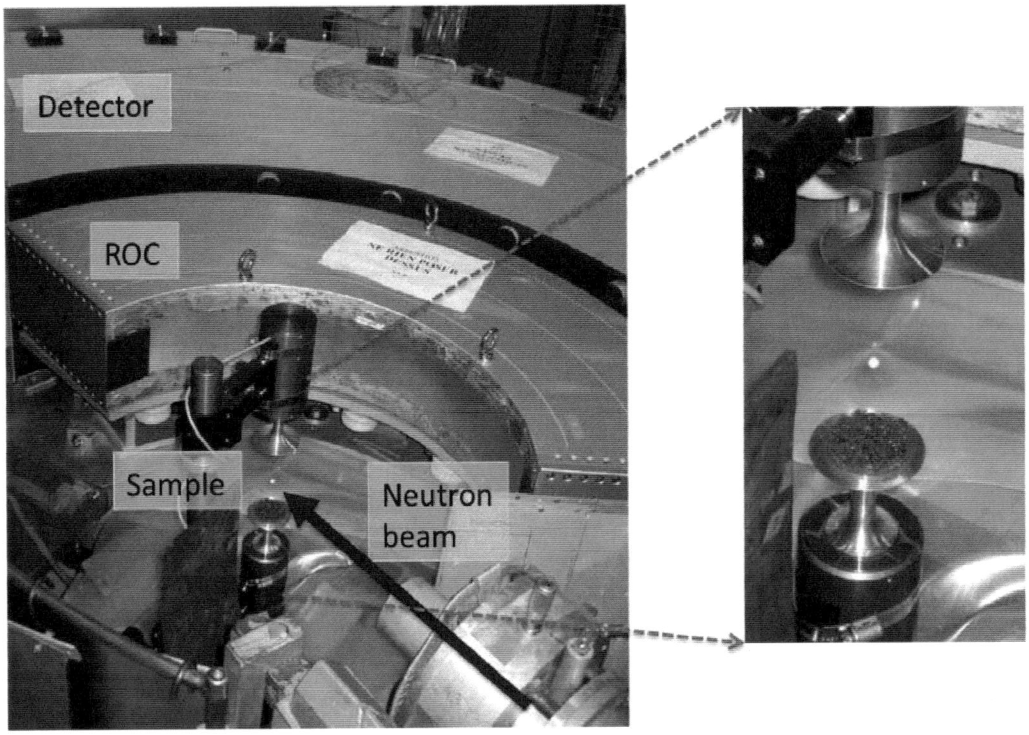

Figure 5. The left picture shows the acoustic levitator placed at the two-axis neutron diffractometer D1B. The neutron beam, sample, ROC and detector are shown. The right picture zooms on the levitated polystyrene sphere.

3.3. Examples

We present in the following a study of three $(MgO)_x(Al_2O_3)_{1-x}$ liquid magnesium aluminates oxides. MgO- and Al_2O_3-containing materials have a strong geophysical and technological importance. They constitute a large fraction of Earth's mantle and they are being considered as inert matrices for nuclear waste confinement, due to their stability under irradiation, high melting point, and high thermal conductivity. In particular, the high melting point (T_m = 2400 K) makes their structural investigation extremely difficult, for that reason the levitation device become essential to carry out the measurements.

Figure 4a shows the $S(Q)$s measured on some magnesium aluminates with $x = 0.33$, 0.50 and 0.75 using the levitation set-up integrated into the D4c diffractometer [14]. With the aerodynamic levitation, statistics are limited by the relatively small sample size that can be levitated. However, reasonable statistics can be obtained. In figure 4b $g(r)$s are presented. Fitting these curves is possible to determine the interatomic distances and the coordination numbers for each pair of atoms for each sample in order to characterize the structure in function of the composition.

Recently it has been tested the feasibility of accomplishing a levitation experiment at D1B using the acoustic levitator, as shown in Fig. 5. D1B is a high flux, medium resolution powder diffractometer at ILL. Although this instrument was designed mainly for determining magnetic structures and phase transition studies on powder samples, several times it has proved its capabilities for studying structural properties of liquids and gases [22, 23, 24, 25]. Due to its geometrical characteristics and set of monochromators, it is possible to obtain in a few minutes a diffraction pattern over a Q range of 8.82 Å^{-1}.

For this test, polystyrene samples were placed in the center of D1B neutron beam while levitating (Fig. 5). As usual, a radial oscillating collimator was used to reduce the background produce by the sample's surroundings. Solid samples are stable for several hours and it is possible to place few samples in the beam along the vertical axis of the levitator, using the nodes of the standing wave. We have also levitated drops of light and heavy water and collected the corresponding diffractograms; in this case, some sample is lost due to the slow evaporation at ambient conditions. In the case of polycrystalline powder samples, the main problem is to finding a way to keep the powder in the beam. In this test we use Si powder as sample, and this powder was wrapped in Al foil. The Fig. 6 illustrates the diffraction pattern of 17 mg of Si wrapped in 12 mg of Al foil collected along 60 minutes. One can observe the diffracted peaks arising from the Si and Al, which are perfectly suitable for Rietveld refinement.

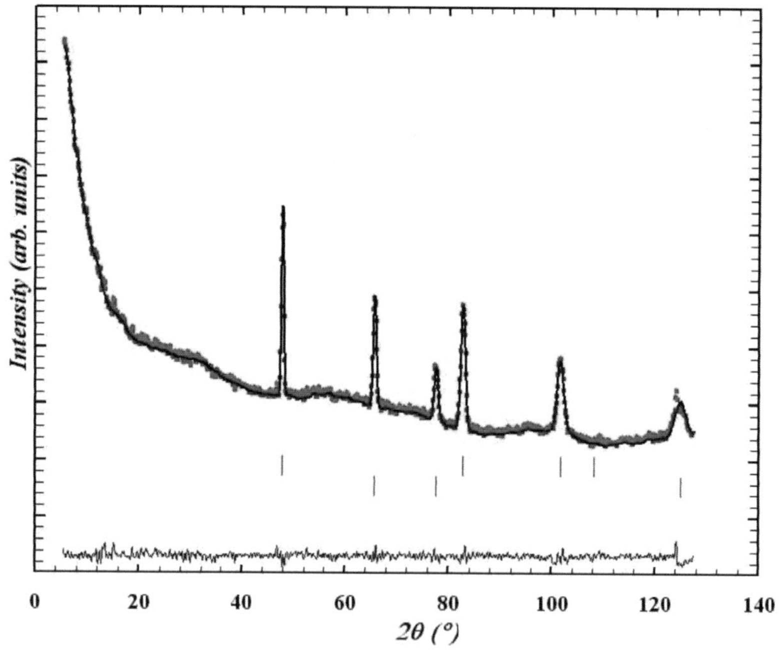

Figure 6. (colour online) Observed (dot) and calculated (solid line) diffraction pattern of 17 mg Si wrapped in a 12 mg Al foil, after a data collection of one hour at a wavelength of 2.52 Å. The vertical bars indicate the position of the Bragg diffraction reflections; first row corresponds to the Bragg reflections belonging to Si and the second to those of the Al. The difference between observed and calculated is depicted on the bottom of the figure.

4. Conclusions

The sample environment is a critical component of research programs in advanced materials, geological systems, biology, and energy-related research and development. Levitation techniques have an increasing interest in recent years as important tools for scattering experiments. The main reasons for this interest are the low background signal, the purity of the sample and the possibility of performing experiments with small amounts of samples. In particular, we have shown that diffraction experiments are possible on D4, D1B and D22 with samples of only a few tens of mg. Among the different containerless methods, aerodynamic levitation combined with laser heating has proved to be a powerful and versatile technique for studying the structure, dynamics, and macroscopic properties of high-temperature liquids including insulating compounds (glasses, ceramics, and oxides) and metallic materials. The use this techniques is mainly adapted for studies of materials above the melting point and in the supercooled state condition maintaining a high degree of control in temperature. In particular, the atomic structure of liquid materials can be determined by measurements of the structure factor $S(Q)$ and calculation of the corresponding pair correlation function $g(r)$, as reported in this paper. The acoustic levitation is originally designed for the investigation of supercooled liquids and aqueous solutions in the temperature range from about -40 C to $+40$ C. This temperature range covers the supercooled region for water and many salt and biological solutions. By eliminating contact with container walls, sources of heterogeneous nucleation are avoided and it often becomes possible to cool liquids substantially below their equilibrium melting point. Here we show that this technique can be also adopted for powder diffraction with a small amount of sample in the beam obtaining good data quality necessary to make the data refinement. In particular, we have shown that diffraction experiments are possible on D4, D1B and D22 with samples of only a few tens of mg. The high adaptability on different neutron instruments of the levitation devices presented in this article offers the possibility to combine different neutron techniques. This is a very important point for an exhaustive analysis of materials based on multi-techniques and multi-scale approach.

References

[1] Price D L 2010 *High-temperature Levitated Materials* (Cambridge: Cambridge University Press).
[2] Hennet L, Thiaudière D, Gailhanou M, Landron C, Coutures J P and Price D L 2002 *Rev. Sci. Instr.* **73** 124
[3] Greaves G N, Wilding M C, Kargl F, Hennet L, Majérus O 2008 *Adv. Mat. Res.* **39-40** 3
[4] Skinner L B, Barnes A C, Salmon P S and Crichton W A 2008 *J. Phys. Condens. Matter* **20** 205103
[5] Holland-Moritz D, Schenk T, Convert P, Hansen T and Herlach D M 2005 *Meas. Sci. Technol.* **16** 372
[6] Jacobs G, Egry I, Maier K, Platzek D, Reske J and Frahm R, 1996 *Rev. Sci. Instrum.* **67** 3683
[7] Paradis P F, Ishikawa T, Yu J and Yoda S 2001 *Rev. Sci. Instrum.* **72** 2811
[8] Gangopadhyay A K, Lee G W, Kelton K F, Rogers J R, Goldman A I, Robinson D S, Rathz T J and Hyers R W, 2005 *Rev. Sci. Instrum.* **76** 073901
[9] Trinth E H 1985 *Rev. Sci. Instrum.* **56** 2059
[10] Granier J and Potard C 1987 *Proc. 6th Eur. Symp. Mater. Sci. Micrograv.* ESA SP-256
[11] Haumesser P H, Garand J P, Bancillon J, Daniel M, Campbell I and Jackson P 2002 *Int. J. Thermophys.* **23** 1217
[12] Winborne D A, Nordine P C, Rosner D E and Marley N F 1976 *Metall. Trans. B* **7** 711
[13] Coutures J P, Massiot D, Bessada C, Echegut P, Rifflet J C and Taulelle F 1990 *C. R. Acad. Sci.* **310** 1041
[14] Hennet L, Pozdnyakova I, Bytchkov A, Price D L , Greaves G N, Wilding M C, Fearn S, Martin C M, Thiaudière D, Bérar J-F, Boudet N and Saboungi M-L 2007 *J. Chem. Phys.* **126** 074906
[15] Hennet L, Pozdnyakova I, Cristiglio V, Cuello G J, Jahn S, Krishnan S, Saboungi M-L and Price D L 2007 *J. Phys. Cond. Mat.* **19** 455210
[16] Hennet L, Pozdnyakova I, Bytchkov A, Cristiglio V, Palleau P, Fischer H E, Cuello G J, Johnson M R, Melin P, Zanghi D, Brassamin S, Brun J-F, Price D L and Saboungi M- L 2006 *Rev. Scient. Instrum.* **77** 053903
[17] Landron C, Hennet L, Coutures J P, Jenkins T, Aletru C, Greaves N G, Soper A K and Derbyshire G 2000 *Rev. Sci. Instrum.* **71** 1745
[18] Meyer A, Stüber S, Holland-Moritz D, Heinen O and Unruh T 2008 *Phys Rev. B* **77** 092201
[19] Weber J K R, Rey C A, Neuefeind J and Benmore C J, 2009 *Rev. Sci. Instrum.* **80** 083904

[20] Cristiglio V 2008, Structural characterisation of molten aluminates. Experiments and modelling, PhD Thesis, University of Orléans

[21] Fischer H E, Cuello G J, Palleau P, Feltin D, Barnes A C, Badyal Y S and Simonson J M 2002 *Appl. Phys. A: Material Sci. Process.* **74** S160

[22] Pothoczki S, Ottochian A, Rovira-Esteva M, Pardo L C, Tamarit J Ll and Cuello G J 2012 *Phys. Rev. B* **85** 014202

[23] Sanz A, Nogales A, Puente-Orench I, Jiménez-Ruiz M and Ezquerra T A 2011*Phys. Rev. Lett.* **107** 025502

[24] Pardo L C, Rovira-Esteva M, Tamarit J Ll, Veglio N, Bermejo F J and Cuello G J 2010 in *Metastable systems under pressure*, eds S. Rzoska, A. Drozd-Rzoska and V. Mazur (Dordrecht: Springer) p 79

[25] Negrier P, Barrio M, Tamarit J Ll, Pardo L C and Mondieig D (2012) *Crystal Growth Design* **12** 1513

The new powder diffractometer D1B of the Institut Laue Langevin

I Puente Orench[*,1,2], J.F. Clergeau[2], S. Martínez[1,2], M. Olmos[1,2], O. Fabelo[1,2], J. Campo[1]

[1]Instituto de Ciencia de Materiales de Aragón, (CSIC-Universidad de Zaragoza), c/ Pedro Cerbuna 12, 50009, Zaragoza. Spain.
[2]Institut Laue Langevin, 6 rue Jules Horowitz, 38000 Grenoble, France.

E-mail: puenteorench@ill.eu

Abstract. D1B is a medium resolution high flux powder diffractometer located at the Institut Laue Langevin, ILL. D1B a suitable instrument for studying a large variety of polycrystalline materials. D1B runs since 1998 as a CRG (collaborating research group) instrument, being exploited by the CNRS (Centre National de la Recherche Scientifique, France) and CSIC (Consejo Superior de Investigaciones Científicas, Spain). In 2008 the Spanish CRG started an updating program which included a new detector and a radial oscillating collimator (ROC). The detector, which has a sensitive height of 100mm, covers an angular range of 128°. Its 1280 gold wires provide a neutron detection point every 0.1°. The ROC is made of 198 gadolinium-based absorbing collimation blades, regular placed every 0.67°. Here the present characteristics of D1B are reviewed and the different experimental performances will be presented.

1. Introduction

The neutron diffractometer D1B at the Institut Laue Langevin (ILL), was built in the 70´s. It was designed mainly to study magnetic structures; however its characteristics make possible to investigate a vast variety of scientific problems, such as phase transitions by thermo-diffraction experiments, time-resolved or *"in situ"* studies like crystallization experiments, or to perform texture analysis. Since 1998 D1B is a CRG (collaborating research group) instrument, handled by the CNRS (Centre National de la Recherche Scientifique, France) and CSIC (Consejo Superior de Investigaciones Científicas, Spain). In the last years the Spanish CRG has undergone some technological updating of the instrument, mainly a new position sensitive detector (PSD) and a radial oscillating collimator (ROC) have been built, commissioned and now are at the service of all users of D1B

2. Neutron delivery

The thermal curved guide H22 delivers neutrons to three instruments D1B, SALSA and VIVALDI, being D1B the first of them up stream (see Figure 1). The monochomator´s casemate of D1B accommodates two monochromators mounted on a set of motors. Thus, it is possible to change from one to the other in few minutes. Neutrons are delivered to the sample with a take off angle of 44.22°. On one hand, three highly oriented pyrolytic graphite (HOPG) crystals supply a neutron beam of 2.52Å focalized on the sample, with a flux of $5.6x10^6 n/cm^2s$; on the other a germanium single crystal

[*] To whom any correspondence should be addressed.

provides a neutron beam of 1.28 Å, however the neutron flux at the sample position is ten times lower. Downstream of the monochromators there is a set of movable graphite filters used after the HOPG monochromator which reduce the second harmonics up to 8/10000.

The already monochromated neutrons pass through an evacuated tube where there are 5 diaphragms which collimate the beam. A set of adjustable horizontal-vertical slits, permit to accurately define the size of the beam on the sample.

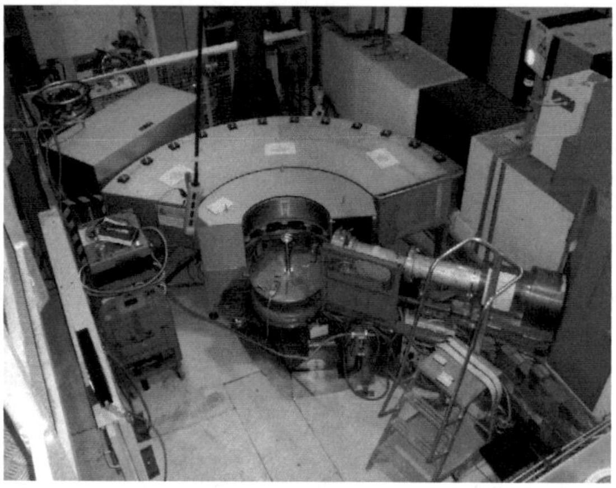

Figure 1.Top view of the D1B neutron diffractometer at ILL.

3. Neutron detection

Since June 2011, it is in operation a new detector at D1B. After more than thirty years working, the old detector of D1B was in a big risk of failure. It was in 2008 when the Spanish part of the CRG D1B decided to start a project for building a new detector for D1B. Thanks to funding from the Ministerio de Ciencia e Innovación (present Ministerio de Economía y Competitividad) in less than three years it has been possible to replace the detector of D1B. The new large PSD for D1B was designed and constructed in a fruitful collaboration between the Spanish CRG and the ILL.

The current detector of D1B covers 0.15 steradians. 1280 W/Re gold plated wires placed every 0.1° inside a sealed gas chamber are the responsible of detecting the neutrons scattered by the sample. The gas filling of 5bars of ^3He and 1bars of CF_4 provides a detection efficiency of 80% at 2.52 Å and 60% at 1.28 Å. The PSD is placed onto a vibration isolated table at 1500mm of the sample position.

In Table 1 the main characteristics of the new and old detectors are detailed. Although the geometry remains the same, the improvements are evident, such as the present detector has incremented the 2θ aperture in 48° or the detection efficiency has increased. Summarizing, under the same conditions, the current detector of D1B allows obtaining a larger diffraction pattern in less time than what was with the old one. One can observe this in Figure 2, where are shown the diffraction patterns of a $Na_2Ca_3Al_2F_{14}$ standard sample mounted on a vanadium cylinder sample holder of 5mm in diameter, measured with both detectors at room temperature. There, it is also observable the neutron dead detection zone the old detector had after an electronic failure occurred in 2008.

Table 1. Main characteristic of the old and current detector of D1B.

	New	Old
angular range	128°	80°
height of the sensitive area	100mm	100mm
angular definition	0.1°	0.2°
readout	1280 cells 1280 anodes encoding	400cells 20 anodes and 20 cathodes encoding
radius of curvature	1500mm	1500mm
efficiency: 2.52 Å	80%	60%
1.28 Å	60%	37%

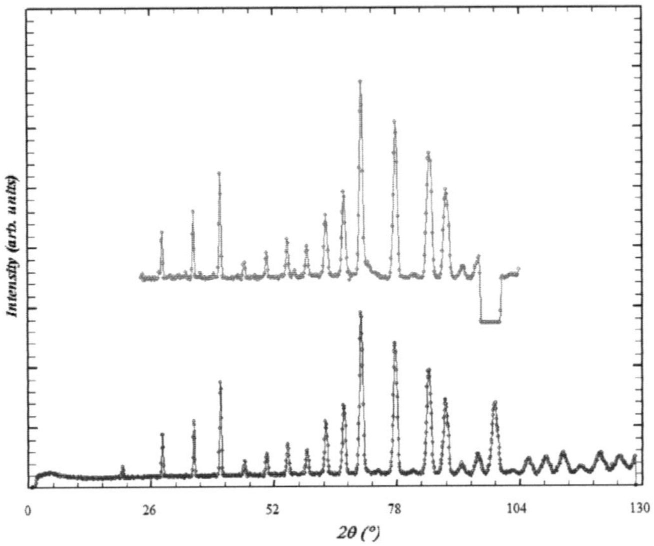

Figure 2. Diffraction pattern at 2.52Å of a ϕ=5mm $Na_2Ca_3Al_2F_{14}$.standard sample. Upper curve: 15minutes acquired with the old detector. Lower curve: 10 minutes acquired with the new detector.

4. Radial oscillating collimator

Many studies accomplished at D1B have the inconvenient that the diffraction peaks coming from the sample are merged with the diffraction arising from the environment. This is the case of cryomagnets, cryofurnaces, or pressure cells. D1B's radial oscillating collimator have been designed and constructed in order to suppress these non-desired scattered neutrons.

The ROC of D1B consist of 198 absorbing collimation blades placed radial in between the sample and the detector every 0.67°. Blades´ dimensions are 350x100x0.1 mm, made of PETP (polyethylene terephthalate) film coated with a neutron absorbing Gd_2O_3 paint. Each blade is under mechanical vertical tension for maintaining the radial structure (see Figure 3). In order to balance the shadows cast by the finite thickness of the absorbing blades, the ROC oscillates at a constant velocity. The oscillation half period is 30s with oscillation amplitude of 6 blade spacing. The dead time of acceleration/deceleration is around 0.5s. The solution adopted at D1B for having such a constant movement is a rotated heart-shape came which transmits the movements tangentially. The ROC has a full width focus of 18.9mm centered on the sample position, i.e. a given point on the detector can therefore see up to 9.4mm to the left and right of the sample's center. In this way neutrons scattered from materials placed beyond will be absorbed by the blades. Unfortunately, some neutrons diffracted from the sample are also blocked by the ROC. For a cylinder sample of 5mm in diameter, 86% of the neutrons scattered by it will arrive to the detector; for a thicker sample of 8mm 78% of neutrons will be transmitted.

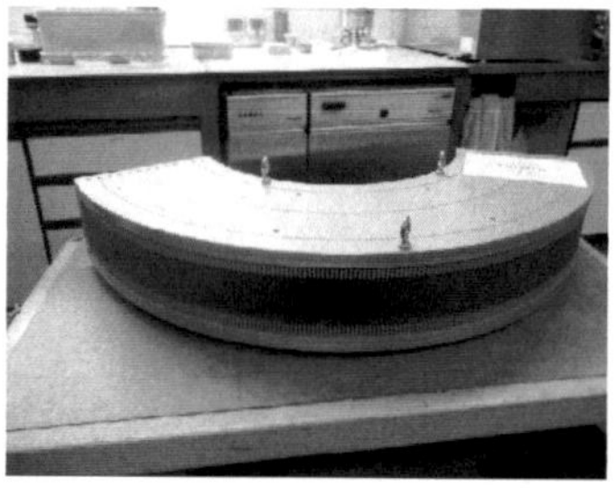

Figure 3. D1B´s ROC radial absorbing blades.

The advantage of using such a device can be observable in Figures 4 y 5. The former illustrates the diffraction pattern of $Na_2Ca_3Al_2F_{14}$.standard sample inside a cryomagnet, where one can hardly recognize the characteristic peaks of this material. The latter shows what it is obtained when the ROC is in place, there all neutrons scattered by the environment have been absorbed. As mentioned above some neutrons diffracted by the sample are absorbed by the ROC, this in principle would increase the acquisition time, however this lost of neutrons arising from the sample are compensated, in most of the cases, with the reduction of background that the ROC causes.

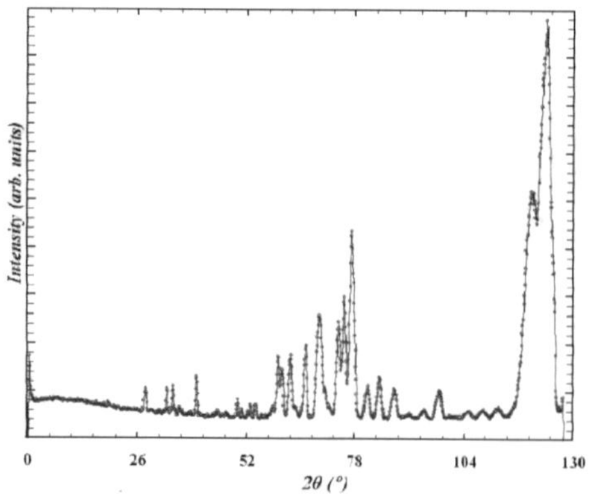

Figure 4 Diffraction pattern at 2.52Å of a ϕ=5mm $Na_2Ca_3Al_2F_{14}$.standard sample measure inside a cryomagnet.

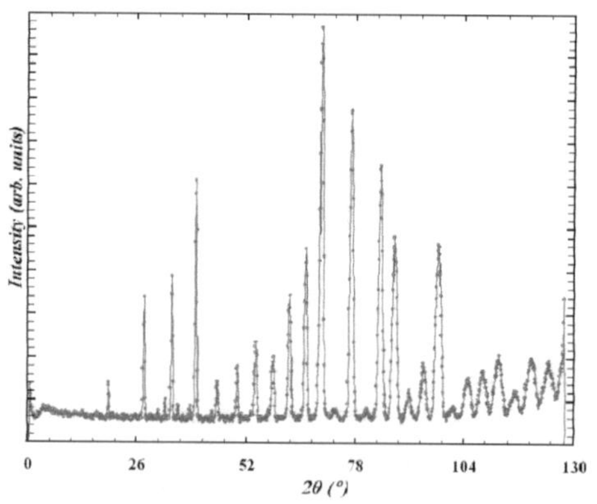

Figure 5. Diffraction pattern at 2.52Å of a ϕ=5mm $Na_2Ca_3Al_2F_{14}$.standard sample measure inside a cryomagnet with the ROC in place.

5. Sample environment

D1B has three dedicated sample environments, an Orange cryostat (2K<T<300K) with a vanadium tail of ϕ=24mm, a furnace (T<1000ºC) with a vanadium resistor ϕ=50mm and a 4-circle Eulerian cradle goniometer for studying textured samples at room temperature.

Moreover, it is also feasible to use the sample environments from the ILL- "pool" like furnaces (T<1600ªC), cryofurnaces (2K<T<550K), cryomagnet (up to 5T), gas and clamp pressure cells or the Paris-Edimburg pressure cell. Recently has been also tested the feasibility of using containerless sample environment under acoustic levitation, with very promising results.

All sample environments at D1B are mounted in a X-Y translation platform (ϕ=700mm) which is on a 360º ω-axis. This makes possible to place in the neutron beam several set-ups.

6. Data acquisition and instrument control

From a standard PC computer running in LINUX, sample conditions and detector acquisitions are controlled. Two programs are offered for instrument control, a command line based: MAD and NOMAD based in a graphical user interface. In addition, a WINDOWS and LINUX workstations are at disposal of the user for reducing and analyzing of D1B´s data, where most used programs like LAMP [1,2] or the FullProf suite [3] can be found.

7. Conclusions

The present performance of D1B, the new detector and the ROC, permits now to face scientific problems which before had to be resolved in other instruments. The increment of the angular definition and the efficiency in neutron detection of the new detector permit to lower the acquisition time of a diffraction pattern in comparison with the old one. These improvements implemented together with the use of the radial oscillating collimator open the possibility of using a vast set of sample environments (cryofurnaces, cryomagnets, etc.) which has a big impact in the feasibility of experiments at D1B.

The ideal situation of D1B in the guide hall of the ILL (high flux and low background) and the upgrades described in this work have pushed D1B to continue being a highly demanded powder diffractometer with a high rate of publications.

Acknowledgments

This work has been funded by the Ministerio de Economía y Competividad of the Spanish Government by means of MCINN-ICTS-2008-36 and ICTS-06-11 projects.

Authors also acknowledge ILL services for the invaluable contribution as well as the members of D1B instrument that along the years have helped to the technological updating of D1B.

[1] Richard, D.; Ferrand, M.; Kearley, G.J. J. Neutron Research 1996, 4, 33-39.

[2] LAMP, the Large Array Manipulation Program. http://www.ill.fr/data_treat/lamp/lamp.html.

[3] Rodríguez-Carvajal, J. *Physica B* **1993**, *55*, 192. The programs of the FullProf Suite and their corresponding documentation can be obtained from the Web at http://www.ill.eu/sites/fullprof/

LAGRANGE - the new neutron vibrational spectrometer at the ILL

Mónica Jiménez-Ruiz, Alexander Ivanov and Stephane Fuard
Institut Laue-Langevin, 6 rue Jules-Horowitz, BP 156, F-38042, Grenoble Cedex
E-mail: jimenez@ill.fr

Abstract. The new neutron vibrational spectrometer LAGRANGE replaces the Be filter spectrometer placed on the hot source at the Institut Laue-Langevin. Lagrange provides much higher sensitivity to extremely small or low scattering samples through substantially increased detector count rate combined with considerably improved energy resolution. It opens new possibilities for the spectroscopic studies of complex materials. The new instrument permits investigation of much smaller samples, inherent in novel materials studies, recording of well-resolved vibration spectra. Lagrange is fully operational for users since 2012.

1. Introduction

Vibrational spectroscopic techniques provides unique information about the high-energy atomic and molecular vibrations in condensed matter systems. The neutron scattering spectroscopy has proven to be a powerful and unique tool for studies of dynamical properties of solids and associated physical and chemical phenomena. Neutrons probe the bulk properties of samples and can easily penetrate through container walls making possible studies of unstable or conditioned materials. The information obtained by Inelastic Neutron Scattering (INS) experiments is complementary to the one obtained by means of infrared and Raman spectroscopy that yield information about lattice vibrations, which produce a variation of the dipole moment or the polarizability of the system respectively. The INS method directly follows the motions of nuclei and, since there are not specific selection rules, the INS spectra provide information about all vibrations of the system. For this reason INS technique is used as an important source of supplementary data for computational techniques dealing with the dynamics of complex systems.

The Lagrange project was launched with the goal to create a new spectrometer for studies of lattice and molecular excitations in the extended energy range up to several-hundred-meV typical for materials containing light chemical elements, in particular hydrogen. Lagrange replaces the Be filter spectrometer placed on the hot neutron spectrometer IN1 at the Institute Laue-Langevin [1] and dedicated to the studies of molecular vibrations in complex systems, such carbon nanotubes [2], zeolites [3] and hydrides [4]. Another instrument dedicated to the molecular vibrational spectroscopy is the high-resolution crystal analyser spectrometer, TOSCA, at ISIS pulsed neutron source [5].

2. IN1-Lagrange design

Lagrange is placed on the hot neutron beamtube H8 at the Institut Laue-Langevin (ILL) and shares this position with the diffractometer for liquids and amorphous systems D4 and with the hot neutron Three-Axis spectrometer IN1. Since Lagrange shares the primary spectrometer with these instruments, it benefits from the recently upgraded double-focussing monochromator with a set of different reflecting planes:

• Cu(220) reflection, with an starting incident energy of $\sim$ 27 meV up to $\sim$ 500 meV, used for the high flux configuration,
• Cu(331) reflection, with an starting incident energy of $\sim$ 65 meV up to $\sim$ 1 V, used for the improved energy resolutions configuration at high energies, and
• Si(311) reflection, with an starting incident energy of $\sim$ 16.5 meV up to $\sim$ 200 meV; and the Si(111) reflection, with an starting incident energy of $\sim$ 5.5 meV up to $\sim$ 20 meV used for the low-energy part of the spectra and that present the advantage of suppressed second order contamination in the monochromatic beam.

With the present reflections, this monochromator supplies at present more than twice increased monochromatic flux in the high neutron energy range available from the hot neutron source of ILL reactor.

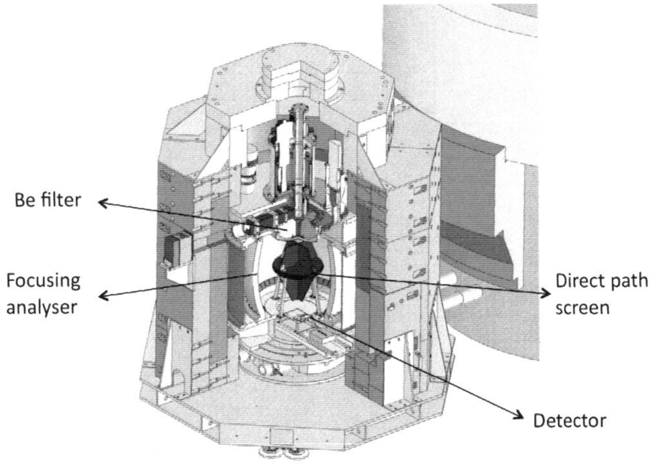

Figure 1. Lagrange design.

The LAGRANGE (LArge GRaphite ANalyser for Genuine Excitations) secondary spectrometer setup (Fig. 1) is based on the space focussing of neutrons scattered by the sample in a very large solid angle, all of them registered with a relatively small single counter (a He3 gas detector). The focussing reflecting surface of 1 m^2 is built around the vertical sample-detector axis from pyrolytic graphite crystals set to reflect neutrons with the fixed average energy of 4.5 meV. The appropriately shaped cooled Beryllium filter is installed right after the sample in order to remove higher-order harmonics in the analyser reflections. It is important that the absolute value of the accepted solid angle ($\sim$ 2.5 steradian) stays among the highest of the most ambitious instrument projects at present. At the same time it is realised in the relatively small characteristic spectrometer volume. The high-energy resolution of the new instrument is attained with a relatively low grade of the pyrolytic graphite. The carefully designed screen of boron-containing absorber is installed on the sample-detector axis in order to suppress the intense elastic scattering from the sample. Further reduction of the instrument background, contaminated with the high neutron energy components, is achieved through a massive polyethylene shielding built around the whole analyser. Both the sample cryostat and the filter cooling circuits are equipped with powerful closed-cycle helium refrigerators. Note that there are no mutually moving parts within the secondary spectrometer which is positioned as a whole around the monochromator in order to record the inelastic scattering spectra changing step-by-step the incident energy similarly to a typical experiment on a three-axis spectrometer.

3. Results

Lagrange has been firstly commissioned during the 2nd reactor cycle of the year 2011. The first inelastic scattering spectrum measured with Lagrange is presented on the Fig.2. We used a sample of 2.5-diiodothiophene, $(CH)_2(CI)_2S$, for characterisation of the instrument performance. This material is known to exhibit very sharp vibration lines at low temperatures so that it can be used as a standard for measurements of the spectrometer resolution. The visible sharp peaks in the energy spectrum on Fig.2 are due to vibrations involving hydrogen atoms, present in the sample in a relatively small quantity: 6 milligram of the partial hydrogen weight. It should be noticed that the collected statistics in the energy interval below 200 meV correspond to some 12-30 seconds of the measuring time per point (the difference is due to energy variation of the incident monochromatic neutron flux) so that this informative energy interval could be scanned in less than 2 hours while the whole scan time for the shown spectrum, up to 600 meV, is about 4 hours.

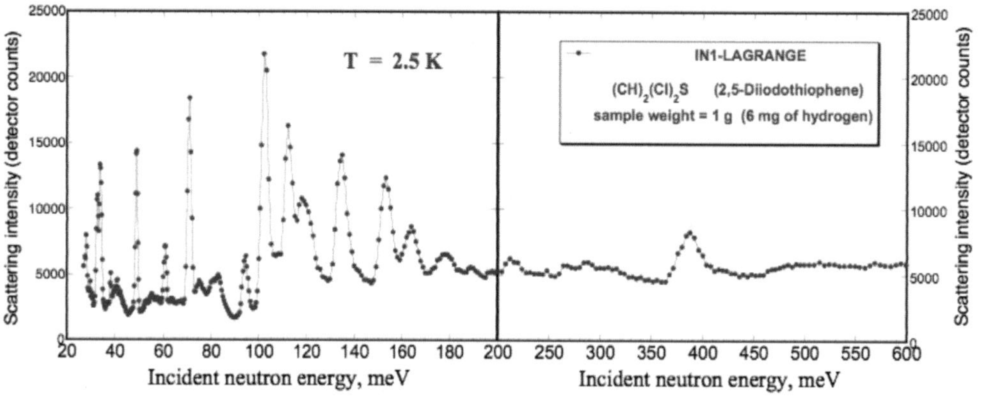

Figure 2. The inelastic scattering spectrum recorded with the Lagrange analyser and the 2D-focusing Cu220 monochromator. The sample of 2.5-Diiodotiophene is cooled down to the base cryostat temperature of 2.5 K. The intensity is as-measured, without any further treatment, no background (cryostat and sample holder) is subtracted.

The energy resolution of Lagrange can be further improved in the broad energy range using of the available Cu331 reflecting plane of the IN1 monochromator, as it is seen from the spectra presented on the Fig.3. Many more fine details of the vibration density of states are revealed in the higher resolution run in the expense of solely $\sim$ 2-fold decrease in the statistics at the same counting time.

The measured line-widths with the 2.5-diiodothiphene sample are presented as points in the Fig.4 which also shows calculated energy resolution curves with different analyser and monochromator settings. We note that similarly to the predictions of the intensity gain,

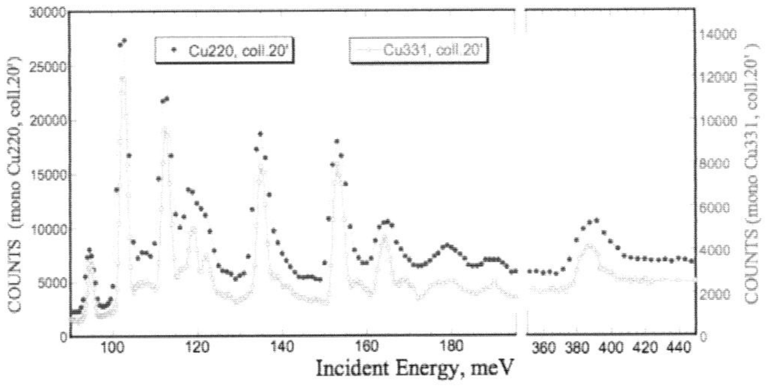

Figure 3. Inelastic scattering spectra (the same sample as on the Figs. 2) measured with different monochromator planes, Cu220 and Cu331. No background is subtracted. The counting time per point was about 20- 30% higher in the case of Cu220 monochromator what corresponds to about a factor of 2 in the intensity decrease at the same counting time for the high-resolution setup.

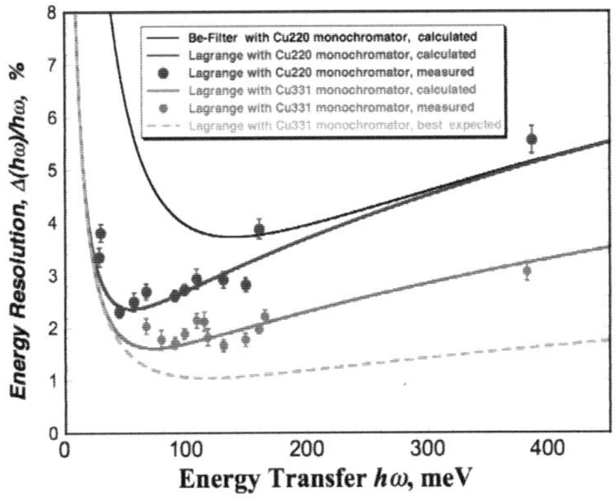

Figure 4. Calculated (lines) and measured (points) relative energy resolution of Lagrange with different primary spectrometer settings. The black line corresponds to the Be-Filter analyser resolution with the Cu220 monochromator, blue line the same in the case of Lagrange, red line Lagrange with the Cu331 monochromator. These calculations are made for the natural collimation of the in-pile beam tube of IN1 and 20 Soller collimator in the monochromatic beam. Points are the peak widths divided by the peak energies as measured in the corresponding conditions with the 2.5-diiodothiophene sample. The lowest broken line represents the best-expected future energy resolution of Lagrange.

the resolution performance is in full accord with what was expected from the pre-project estimations. The best expected energy resolution, of the order of 1% in the most informative energy range 100-200meV, should be achieved with already manufactured 12 and 10 Soller collimators installed before and after monochromator. This ultimately high energy resolution, will bring Lagrange to the top of the high-resolution neutron spectrometers elsewhere and will be available with the future collimator changer installed before the monochromator. The changing over different resolution settings can be very easily done leaving users the flexibility to optimise their measurements with respect to resolution vs intensity choice.

Journal of Physics: Conference Series **549** (2014) 012004 doi:10.1088/1742-6596/549/1/012004

4. Conclusions

The new neutron vibrational spectrometer LAGRANGE has proved as a highly sensitive and a high-resolution instrument. It opens new possibilities for the spectroscopic studies of complex materials. The new instrument permits investigation of much smaller samples, inherent in novel materials studies, recording of well-resolved vibration spectra, a more detailed probing of external parameters like temperature, pressure, chemical composition, magnetic field, time-dependent processes etc. It makes Lagrage highly competitive with the instruments for high-energy neutron spectroscopy elsewhere.

The new instrument provide much higher sensitivity to extremely small or low scattering samples through substantially increased detector count rate combined with considerably improved energy resolution in comparison to the beryllium filter-analyser used in the past on IN1. Lagrange is fully operational for users since 2012 and several improvements are in progress on sample environment (gas injection, heating insert and pressure cells).

References

[1] http://www.ill.eu/instruments-support/instruments-groups/instruments/in1-taslagrange/characteristics/
[2] Rols S, Benes Z, Angleret E, Sauvajol JL, Papanek P, Fischer JE, Coddens G, Schober H and Dianous AJ 2000 *Phys. Rev. Lett.* **85** 5222.
[3] Jobic H 2002 *Current Opinion Sol. State and Mat. Science* **6** 415.
[4] Kolesnikov A *et al* 2002 *Physica B* **316-317** 158-161.
[5] Parker S 2002 *J. Neutron Res.* **10** 173.

The signal of sample environment as a way to measure sample transmission

G J Cuello[1], **M Croibier**[1]

[1] Institut Laue Langevin, 6 rue Jules Horowitz, 38042 Grenoble, France

E-mail: cuello@ill.eu

Abstract. In a recent article, the combined effects of the sample environment and instrument collimation have been described for the dedicated instrument for amorphous systems at ILL (D4) [1]. This undesirable effects on the diffractograms force to make complex and careful absorption corrections, which not always produce good enough results. In such cases, parts of the detector banks must be discarded with the consequent lose of statistics. This experimental problem can be used to obtain the transmission of the sample by an indirect way.

Considering the most generalized case of a cylindrical sample with a cylindrical sample environment (furnace, cryostat, etc), the diffractogram can be described as coming from three different sources: the sample itself, the upstream and downstream sample environment. These sources produce three well-differenced regions on the individual diffractograms observed by one particular detector. These three different measured signals are linear combination of the sample, upstream and downstream intensities produced by each source. In this way, upstream and downstream intensities can be determined, and the experimental transmission of the sample can be obtained. The experimental results corresponding to the standard cryostat are compared with the theoretical calculation of the transmission for a series of samples measured at D4 over the past years. This comparison shows that an undesirable effect observed on the diffractograms could even be useful under some circumstances.

1. Introduction

The transmission of a sample is useful to perform attenuation corrections in scattering experiments. This magnitude gives also the possibility of determining the macroscopic cross section of the sample, which contains the global density of the system.

It is not always easy or possible to perform a direct measurement of the transmission. When it is possible the intensity of the incident beam (no sample in the beam) and the transmitted intensity through the sample should be registered by a detector placed behind the sample position in the beam direction. If the intensity is to high, an attenuator should be placed in front of the detector to avoid any damage. The transmission of the sample is simply calculated as the ratio of the transmitted intensity I and the incident intensity I_0

$$T = I/I_0, \tag{1}$$

where a slab geometry is assumed.

In the case where a sample environment is necessary to perform the experiment, the walls of this device (cryostat, furnace, etc) will be additional sources of radiation being registered by the detector. Usually these walls can not be neither too far from nor too close to the sample position, *i.e.* at some centimeters of the sample. In this way, the intensity scattered by the walls placed upstream and downstream are useful to determine the transmission of the sample [1], without placing a detector in the direct beam, just a detector located at a wide angle position is necessary. The problem is that without collimation in front of the detector, the signals coming from the three sources will be registered together, avoiding in this way any determination of the transmission. The D4 collimation allows this kind of measurements, as it will be shown below.

2. Geometrical effects

The cylindrical symmetry of a two axis diffractometer advice the use of cylindrical samples and environments. Thus the sample environment (a cryostat, a furnace or others) can be represented by a cylindrical screen around and centered at the sample position (see Fig. 1 taken from [1]).

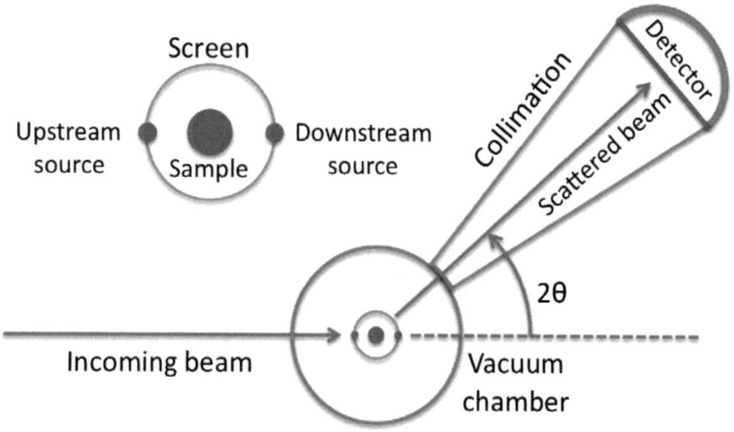

Figure 1. Sketch of the diffractometer device.

Now we focus the discussion on a particular neutron diffractometer having a series of detector banks, each one with a collimation [2]. This collimation is designed in order to register the sample signal with the whole surface of the detector, but tight tight to sufficiently reduce the background. What happens with this collimation is that the central part of the detector sees the sample plus the two extra sources, upstream and downstream, coming from the sample environment. The low-angle part of the detector can only see the upstream signal plus the sample, while the high-angle part can see the sample plus the downstream signal. Figure 2 shows a typical diffractogram, where we can clearly identify the observed signal in the high-, mid- and low-angle regions as I_1, I_2 and I_3, respectively (where $I_3 < I_1 < I_2$). I_3 is smaller than the others due to the attenuation of the sample. In this way, the relationship between these observed intensities and the point sources is

$$\left\{ \begin{array}{lcl} I_1 & = & I_\mathrm{u} + I_\mathrm{s} \\ I_2 & = & I_\mathrm{u} + I_\mathrm{s} + I_\mathrm{d} \\ I_3 & = & I_\mathrm{s} + I_\mathrm{d} \end{array} \right. , \qquad (2)$$

or

$$\left\{ \begin{array}{lcl} I_\mathrm{u} & = & I_2 - I_3 \\ I_\mathrm{s} & = & I_1 - I_2 + I_3 \\ I_\mathrm{d} & = & I_2 - I_1 \end{array} \right. , \qquad (3)$$

where I_s, I_u and I_d are the intensities scattered by the sample, upstream and downstream sources, respectively. These expressions allow the calculation of the experimental sample transmission T_s simply using the Eq. (1) where I_d and I_u are I and I_0, respectively.

We can also obtain a theoretical expression for the transmission of a cylindrical sample of radius R,

$$T_c = \exp(-\frac{\pi}{2}\, n\, \sigma\, R), \tag{4}$$

where n is the atomic density of the sample (number of scatterers per unit volume) and σ is the total (scattering plus absorption) neutron cross section. In the evaluation of the absorption cross section, we use the wavelength-dependent expression

$$\sigma_{\mathrm{abs},\lambda} = \frac{\lambda}{\lambda_{\mathrm{th}}}\, \sigma_{\mathrm{abs}}, \tag{5}$$

where σ_{abs} and λ_{th} are the absorption cross-section and wavelength for thermal neutrons, respectively.

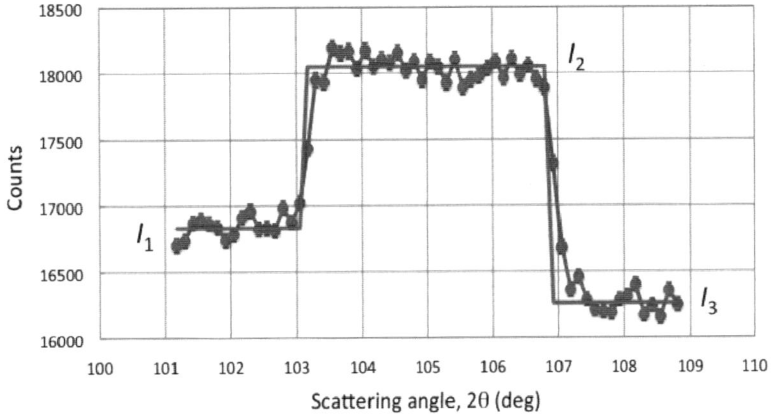

Figure 2. Typical shape of the registered signal on a given detector. The three levels of intensity corresponding to the low- (I_1), mid- (I_2) and high-angle (I_3) regions are shown.

3. Experiments

The diffraction experiments were performed at the two-axis diffractometer D4, ILL, Grenoble, France [2]. We analyzed two independent sets of experiments on a cylindrical vanadium sample of 6.37 mm diameter placed at the center of a cryostat tail also made of vanadium (50 mm inner-diameter and 0.45 mm thick). The incident neutron wavelength was of 0.5 Å and both experiments were performed at room temperature.

Data sets for each experiment correspond to different positions of the detection ensemble constituted of 9 banks of detectors, each one with 64 detection cells. The signal of the 9 banks are independently registered, but only those detectors clearly affected by the shadow effect were used in the evaluation of the three levels of intensity (I_1, I_2 and I_3). This shadow effect (Fig. 2) is maximum for those detectors placed near a scattering angle of 90°. In this way, for the first experiment we could collect 49 independent set of data, while for the second we collected 102 independent data.

For each set of data, constants were fitted to the three observed levels of intensity I_1, I_2 and I_3. These three constants are the input values for Eq. (3), allowing the obtention of the experimental upstream and downstream intensities.

Journal of Physics: Conference Series **549** (2014) 012005 doi:10.1088/1742-6596/549/1/012005

4. Results and discussion

For vanadium, the scattering cross section is 5.1 b and the absorption cross section for thermal neutrons (1.8 Å) is 5.08 b [3]; the absorption cross section corresponding to the wavelength of 0.5 Å is 1.41 b, giving a total cross section of 6.51 b. The vanadium density is $\rho = 6.11$ g/cm^3, which corresponds to an atomic density of $n = 0.07223$ Å^{-3} ($n = \rho/M_{\text{mol}}\, N_{\text{A}}$, where N_{A} is the Avogadro's number and $M_{\text{mol}} = 50.9415$ g/mol is the molar mass). Using these values and the Eq. (4), the theoretical value for the sample transmission is 0.791 ± 0.003.

From the two sets of experimental data, and using Eqs. (1) and (3), we obtain the experimental transmission of the sample 0.75 ± 0.06 and 0.73 ± 0.13. These values compares very well with the theoretical one, confirming the validity of the proposed method.

It is worth noticing that the experimental values are slightly smaller than the theoretical one. This could be explained by the approximations of point sources, made to deduce the relationship between the sources and the observed intensities (Eq. (2)). In fact, the neutron beam is bigger than the sample diameter; in this case, the beam width was of 11 mm meaning that only 55% of the incident beam will pass through the sample. Consequently, part of the beam scattered by the downstream source has not been attenuated by the sample. Taking into account the focalization of the neutron beam on the sample position, the scattered intensity by the downstream source should be reduced by 43%. Thus, the new transmission can be obtained multiplying the previous experimental values by a factor of $0.45/0.43 = 1.05$, which yields to the following values for the experimental transmissions: 0.79 ± 0.06 and 0.77 ± 0.13, much closer to the theoretical value.

There is an experimental way to take into account this background effect. Replacing the sample by an absorbent, the scattered beam not attenuated by the sample can be directly measured. The absorbent sample depends on the energy of the incident neutrons, being Boron the best one for hot neutrons and Cadmium for thermal neutrons. This background could be subtracted from the experimental data and the effect will be more important for the I_{u} intensity than for the I_{d}, which naturally will produce a higher transmission value.

This method could be applied for all the working wavelengths on the instrument, because no influence of the neutron wavelength is expected, except by the known dependence of the absorption cross section with the neutron energy (Eq. (5)).

5. Conclusions

The transmission of a known cylindrical sample has been theoretically calculated using the its density and macroscopic total cross section. This value is in a very good agreement with two independent experimental values obtained applying the method proposed in this article. The method consists in using the scattered signals from the upstream and downstream sample environment as incident and transmitted intensities. Thanks to the instrument collimation, these signals are easily observed on the detection banks as a double step function (Fig. 2). A priori this is an undesirable (but correctable) effect observed on the diffractograms; however it has been demonstrated that it could be useful under some circumstances.

References

[1] Cuello N and Cuello G J 2012 *J. Phys.: Conf. Series* **340**, 012023.

[2] Fischer H E, Cuello G J, Palleau P, Feltin D, Barnes A C, Badyal Y S and Simonson J M 2002 *Appl. Phys. A: Material Sci. Process.* **74** S160.

[3] Sears V F 1992 *Neutron News* **3** 26.

Journal of Physics: Conference Series **549** (2014) 012006 doi:10.1088/1742-6596/549/1/012006

Device and software used to carry out Cyclic Neutron Activation Analysis

M P Castro-García[1], M A Rey-Ronco[2] and TAlonso-Sánchez[3]

[1,3] Departamento de Explotación y Prospección de Minas, Universidad de Oviedo, Oviedo 33004, Spain.
[2] Departamento de Energía, Universidad de Oviedo, Oviedo 33004, Spain.

E-mail: castromaria@uniovi.es

Abstract. This paper discusses the device and software used to carry out Cyclic Neutron Activation Analysis (CNAA). The aim of this investigation is defining through this device the fluorite content present on different samples from fluorspar concentration plant through the DGNAA (Delayed Gamma Neutron Activation Analysis) method. This device is made of americium-beryllium neutron source, NaI (2"x2") and BGO (2"x2") gamma rays detectors, multichannel and an automatic mechanism which moves the samples from activation and reading position. This mechanism is controlled by a software which allows moving the samples precisely and in a safe way (~ms), which it is very useful when the radioactive isotopes have to be detected with a half time less than 8s.

Introduction

Neutron activation consists of bombarding a sample with neutrons for a given time, and subsequently recording or counting the gamma spectrum that is produced during a second specific period of time. In cyclic neutron activation the bombardment and counting phases are repeated for a number of times called a cycle. The number of cycles, the activation or bombardment time, and the counting time are previously determined mathematically, in order to achieve an optimal analysis. These parameters depend, among other things, on the element that one wishes to analyze [1, 2].

The basic components for the device that are needed in order to carry out cyclic neutron activation are as follows: a neutron source, multichannel and gamma ray detectors, a mechanism for transferring the sample, and an enclosure to house the source and the routes along which the sample is to move. The specific characteristics depend on the neutron reaction that will be detected.

In our case, neutron activation is carried out in order to analyze the fluorite content of samples taken from different parts of a concentration plant, using the Delayed Gamma Neutron Activation Analysis (DGNAA) method. The mining sample is fluorspar, which contains other minerals in addition to fluorite, such as silica, barite, iron oxides and silicates.

The fluorite grade in a sample is directly proportional to its fluorine grade, which is, in turn, directly proportional to the quantity of ^{16}N, produced in the reaction $^{19}F(n,\alpha)^{16}N$ [3, 4]. The sample consists of 350 g of the product, which is our case is dry ore with a granulation of less than 100 μm. The device used in neutron activation experiments is conditioned by the characteristics of the ^{16}N that one wishes to detect, and by the cross-section of the reaction.

In addition, as can be seen in Figure 1, the nuclear reaction under study has a high cross-section (on the order of 0.2 barns), whereas the energy of the neutrons hitting the sample is between 3 and 8 MeV.

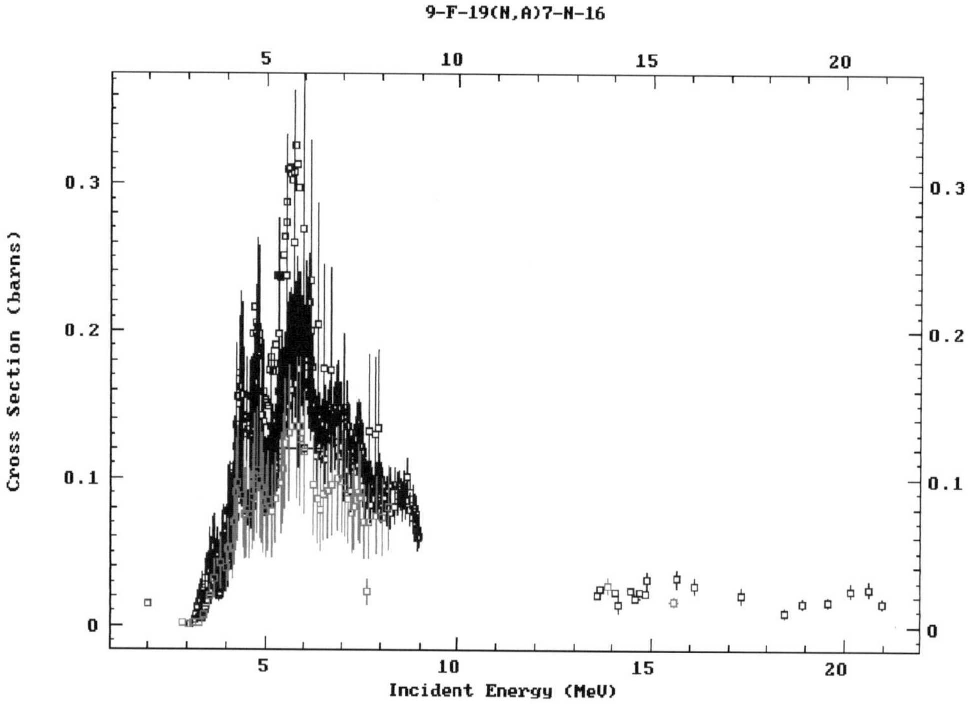

Figure 1. Cross-section of the reaction $^{19}F(n,\alpha)^{16}N$.

Furthermore, the ^{16}N has a half-life of 7.13 s, and emits high-energy gamma rays (6128 keV) as it decays.

These factors necessarily condition the various elements that make up the device needed to carry out the activation.

The neutron source that emits neutrons whose energy is between 3 and 8 MeV is an isotopic americium-beryllium source. A source that produced neutrons with a higher energy level (a neutron generator, for example), would cause the reaction $^{16}O(n,p)^{16}N$. This would bring about ^{16}N which would not have originated from the ^{19}F, and so the concentration of ^{16}N present after irradiation and the ^{19}F present in the sample would no longer be proportional. Because of this, the device is built with an americium-beryllium neutron source, with an activity of 1 Ci.

Of the different types of gamma ray detectors available, we have tested NaI and BGO detectors. Through experimentation we have found that the latter offer better resolution at high gamma energy levels (6128 KeV), such as those emitted by ^{16}N. More specifically, we have used an Osprey multichannel, which behaves better at these energy levels. The detector communicates with the computer through the program Gamma Acquisition & Analysis.

^{16}N's short half-life (7.13 s) means that the time it takes to transfer the sample from the irradiation position to the gamma ray counting station must be as short as possible. Also, the detector should be far enough away from the source to minimize the background spectrum received from the source. To this end, a borated paraffin and lead barrier should be placed in the space between the source and the detector. As a compromise between these two factors, it has been determined that the distance between the detector and the neutron source should be 700 mm.

The mathematical optimization of the process has shown that the time for irradiating the sample and the time for counting the gamma rays emitted by the ^{16}N should be on the order of 12 s each. This being the case, a high degree of control and precision is needed, not only in terms of the transfer time, but also in terms of the time that the sample stays in front of the detector or the source, as well as the exact positions that the sample is to occupy when facing the neutron source or the BGO detector.

It has been shown that over a single activation cycle, the concentration of ^{16}N produced is very low, and that the subsequent gamma ray count is very small [5]. This is overcome by employing various activation-counting cycles, in a process known as cyclic activation (CNAA) [6]. There are different ways to carry out cyclic activation, among them those discussed in the article published in the SETN-2012 conference titled, "New models for carrying out cyclic neutron activation. Discussion of theoretical response and comparison with experimental results," which we have called *sym cyclic activation* and *asym cyclic activation*.

This type of procedure requires a sample transfer system that can handle a relatively high number of stop-and-go cycles, and is able to follow a pattern of acceleration–stop–reverse acceleration, in which acceleration and overclocking can be significant. Due both to this, and to the mass of the sample and its container, the device must be built with a durable structure that is able to withstand the stress of transferring the sample. In this case we used a Servotube linear motor joined to the structure itself, for which the movements, speeds, acceleration and overclocking of the mobile unit can be defined through a software application. The mobile unit is joined to the sample-container unit. Figure 2 shows the complete device, while Figure 3 shows the Servotube linear motor that was used.

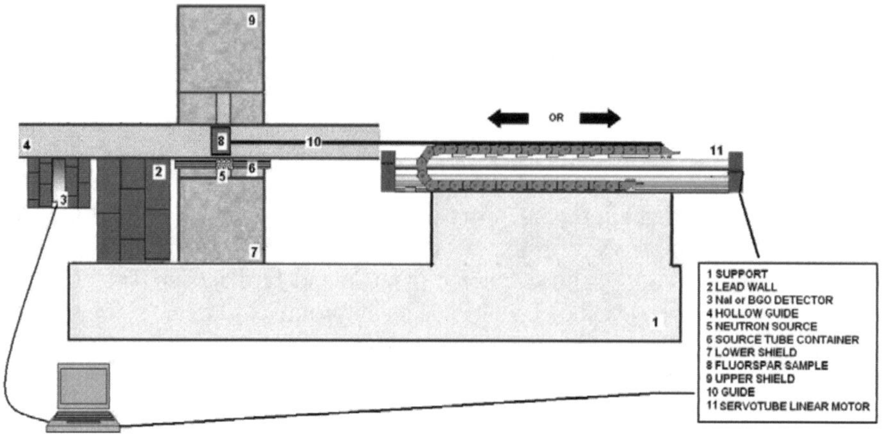

Figure 2. Device for carrying out cyclic activation.

The first step in the cyclic neutron activation procedure is to set a number of parameters: activation, counting and transfer time for the sample, the number of cycles, the detector's calibration parameters, and start and stop parameters for spectrum acquisition. The linear motor and the detector's software must be controlled simultaneously. For this purpose a software program has been developed which synchronically controls the linear motor and the gamma spectrum acquisition's respective programs. It can thus tell the linear motor to transfer the sample to the exact point where it is to be bombarded by the neutron source, and then hold that position for a set time.

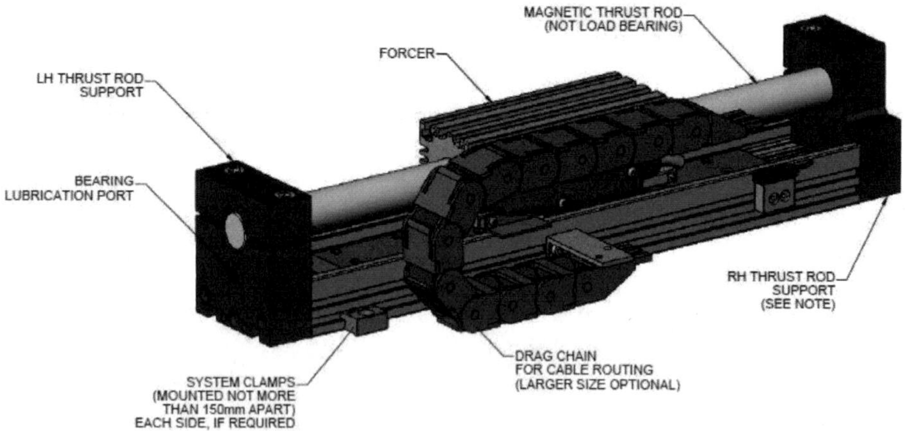

Figure 3. Servotube linear motor (source: www.mecmod.com)

After this it transfers the sample to the exact point where the detector is located, so that it can count the gamma rays emitted for another precise period of time. During this time, the program tells the detector to start reading and recording the accumulated spectrum. The new activation phase begins after the counting time is over, and once spectrum acquisition is complete. The sample's transfer times are set beforehand. A flow chart of the computer program had been designed, in which a process type is first chosen, before moving on to the multichannel selection. After this the control data for the linear motor are entered and the program checks to confirm whether the speed and acceleration results are correct.

Lastly, the program controls when gamma ray spectrum acquisition starts and stops, but only during the time that the sample is placed in front of the detector.

With the device and software discussed in this paper, a variety of types of cyclic neutron activation can be carried out.

Conclusions

This equipment has been designed and made to carry out cyclic neutron activation of fluorite samples from a concentration plant.

It is made of a lineal motor which allows the movement of the samples between the activation position and reading position, in a precise and safe way. Besides this, the device's automation guarantees the detection of ^{16}N which has a half time of 7.13s.

The design incorporates a neutron activation of an activity of 1Ci, a Canberra INA (Tl) detector, 2 in. long and 2 in. in diameter, a Canberra BGO detector, 2 in. long and 2 in. in diameter, an Osprey Multichannel Analyzer (MCA) and a UniSpec Multichannel Analyzer (MCA) with Canberra Genie 2000 Gamma Acquisition & Analysis Software, a Servotube linear motor, 1,130 mm long, 85mm wide and 93 mm tall, with a maximum velocity of 8.7 m/s. This software allows to carry out the activation of the samples.

Acknowledgments

This research was supported and financed by European Regional Development Fund (FEDER 2007-2013) from the Government of the Principality of Asturias (PCTI).

The authors would like to thank the company MINERSA SA, the Government of the Principality of Asturias, and University of Oviedo for their collaboration and financial support throughout the development of this study.

References

[1] Rey-Ronco MA, Alonso-Sanchez T, Castro-Garcia MP 2011 *J Phys Conf Ser* **325** 012025.
[2] Rey-Ronco MA, Alonso-Sanchez T, Castro-Garcia MP 2011 *J Phys Conf Ser* **325** 012021.
[3] Alonso-Sanchez T, Rey-Ronco MA, Castro-Garcia MP 2010 *Int. J. Miner. Process.* **94** 1.
[4] Rey-Ronco MA, Alonso-Sanchez T, Castro-Garcia MP 2010 Nucl. Instrum. *Methods Phys. Res., Sect. B* **268** 2766.
[5] Rey-Ronco MA, Alonso-Sanchez T, Castro-Garcia MP 2010 *J. Math. Chem.* **48** 165.
[6] Rey-Ronco MA, Alonso-Sanchez T, Castro-Garcia MP 2010 *J. Math. Chem.* **50** 325.

Journal of Physics: Conference Series **549** (2014) 012007 doi:10.1088/1742-6596/549/1/012007

New models for carrying out cyclic neutron activation. Discussion of the theoretical response.

M.P Castro-García[1], M.A Rey-Ronco[2], T. Alonso-Sánchez[1]

[1] Universidad de Oviedo. Departamento de Explotación y Prospección de Minas

[2] Universidad de Oviedo. Departamento de Energía

E-mail: castromaria@uniovi.es

Abstract. This paper studies two specific procedures for analyzing mining samples through a neutron activation technique called DGNAA (Delayed Gamma Neutron Activation Analysis). This particular study is part of a broader line of research, whose overall objective is to find the optimal procedure for analyzing the fluorite content of samples taken from different parts of a fluorite concentration plant, using the DGNAA method [1-2]. The mining sample is fluorspar, which contains other minerals in addition to fluorite, such as silica, barite, iron oxides and silicates. The main contribution of the article is the development of a new method for determining the fluorite content in minerals and the increase of sensitivity in respect to the symmetrical method and single-cycle activation.

1. Introduction

Neutron activation consists of bombarding a sample with neutrons, and recording the spectrum of radiation that is produced. In previous articles [3-5], this research team demonstrated that the fluorite grade in a sample is directly proportional to its fluorine grade, which is, in turn, directly proportional to the quantity of ^{16}N, produced in the reaction $^{19}F(n,\alpha)^{16}N$. The team found that there was no interference in either of the following two important aspects: there is no other reaction between the elements in the sample and the neutrons that could produce ^{16}N, and there is no other radioactive product that emits gamma rays at around 6128 keV, which is the energy of the gamma rays emitted by ^{16}N. The procedure followed in prior studies consisted of a single activation process called "monocycle activation," which was made up of two phases: a neutron bombardment phase, and a phase for counting the gamma rays emitted by the sample upon irradiation. However, this procedure yielded a low level of fluorine activation, which in turn led to a poor count of the gamma rays emitted (one count represents a 0.5% of grade mineral) [6].

To overcome this drawback, two theoretical models of cyclic activation have been designed [7], with n activation and counting cycles called "sym cyclic neutron activation" and "asym cyclic neutron activation," the characteristics and results of which are discussed in this article.

2. Theoretical models of cyclic activation

Figures 1 and 2 are graphic representations of the different models: the horizontal axis shows experimental time t in seconds, while the vertical axis shows the evolution of the concentration of the radioactive product, in this case 16N. While the sample is being bombarded with neutrons, the concentration of 16N, g(t), increases according to a known law which is given by the expression:

$$g(t) = k \cdot \left(1 - e^{-\lambda \cdot t}\right).$$

Where $\lambda = \frac{\ln(2)}{T_{1/2}}$, in which $T_{1/2} = 7.13s$ is the half-time of the ^{16}N and k is a parameter that depends on several factors, among them, the concentration in the sample of the element under study (F in this case), the type and intensity of the neutron source, the cross-section of the reaction, or the spatial arrangement of the device's components. While the mineral sample is in front of the neutron source, the concentration increases according to this exponential expression. This time is called "activation time," and it is represented by τ. Once the sample is removed from the radiation source and placed in front of the radiation detector, the 16N concentration diminishes according to a function that depends on the initial concentration, which is $k\cdot(1\text{-}e^{-\lambda\cdot\tau})$. This decay function is:

$$h(t) = k \cdot \left(1 - e^{-\lambda \cdot \tau}\right) \cdot e^{-\lambda \cdot (t-\tau)}$$

The sample is left in front of the gamma ray detector during the so-called "counting time." For the energy interval emitted by 16N, namely around 6000 keV, the reading for a given instant is proportional to the concentration of 16N at that instant. The gamma ray detector logs the spectrum accumulated during this time, and the log for the energy interval mentioned above is proportional to the shaded area in Figure 1, i.e. to the integral of the function h(t) over the period corresponding to the counting time. In this one-cycle study, $A = \int_0^T h(t) \cdot dt$, for each value of T the function A has a maximum that it is produced in an activation time τ, which mathematically it is found when we derive the previous equation and equal it to zero, being T=cte, it was concluded that $\tau_{optimo} = T/2$, the maximum area occurred when the activation and counting times were equal.

The characteristic trait of cyclic activation is that after the first cycle the sample is again irradiated with the neutrons. At the end of the second cycle of irradiation, the total concentration of ^{16}N is the sum of the residual concentration from the first cycle (B2), plus the concentration produced during the second cycle (B1 is kept continuous). In sym cyclic activation, the activation time and counting times are equal for each of the n cycles. Under these circumstances, the area for the total experimental time T and n cycles is proportional to $A_{max}^{*n\ cycles}(s)$ which is defined by:

$$A_{max}^{n,cycles} = \frac{n \cdot k}{\lambda} \cdot \left(1 - e^{-\lambda \cdot T/2 \cdot n}\right)^2$$

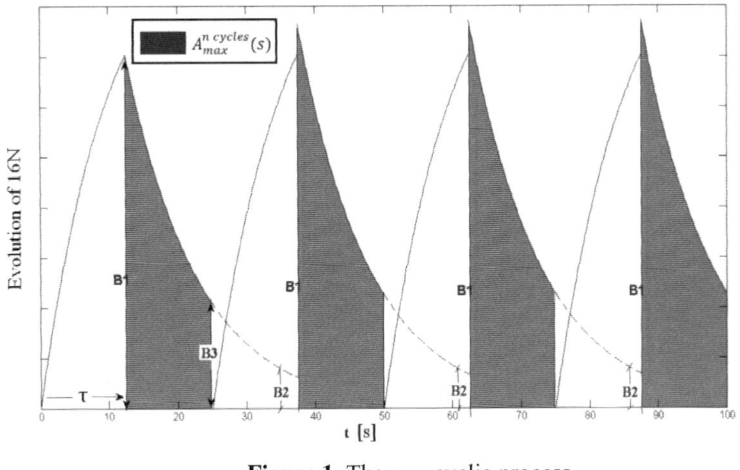

Figure 1. The *sym* cyclic process.

The gain G achieved by sym cyclic activation over single-cycle activation, for the same total experimental time T, depends on both T and n, and is defined by the expression:

$G = n \cdot \frac{\left(1-e^{-\lambda \cdot T/_{2 \cdot n}}\right)^2}{\left(1-e^{-\lambda \cdot T/_{2}}\right)^2}$ As such, for each experimental time T there is an n at which G is maximized, which it is obtained when $\frac{dG\,(n)}{dn} = 0$. As such, for each experimental time T there is an n at which G is maximized. For example, for T=100s, cyclic activation's gain is maximized at n=4 cycles, and the value of this gain is 2.010. Table 1 shows the theoretical gains for different values of T and n, which are obtained by replacement in the previous equation.

	Theoretical Vs n=1			
	Time of process, T			
n	***300***	***200***	***160***	***100***
2	1.997	1.969	1.921	1.690
3	2.954	2.770	2.570	1.961
4	3.794	3.327	2.940	2.010
5	4.473	3.672	3.114	1.963

Table 1. Theoretical gain of cyclic activation over single-cycle activation.

The process called asym cyclic activation is characterized by having an initial activation cycle with a higher activation time than subsequent cycles. It could be said that the first activation cycle is made up of two parts: a "normal" activation time ta, where the 16N concentration is B3, and it is equal to the reading time tl. In this moment, the concentration is reduced from B1 to B2, and it is repeated in all the cycles, plus an "extra" activation time tx, which pushes the concentrations of ^{16}N up to a higher range from B3 to B1, than in sym cyclic activation , as can be observed in Figure 2.

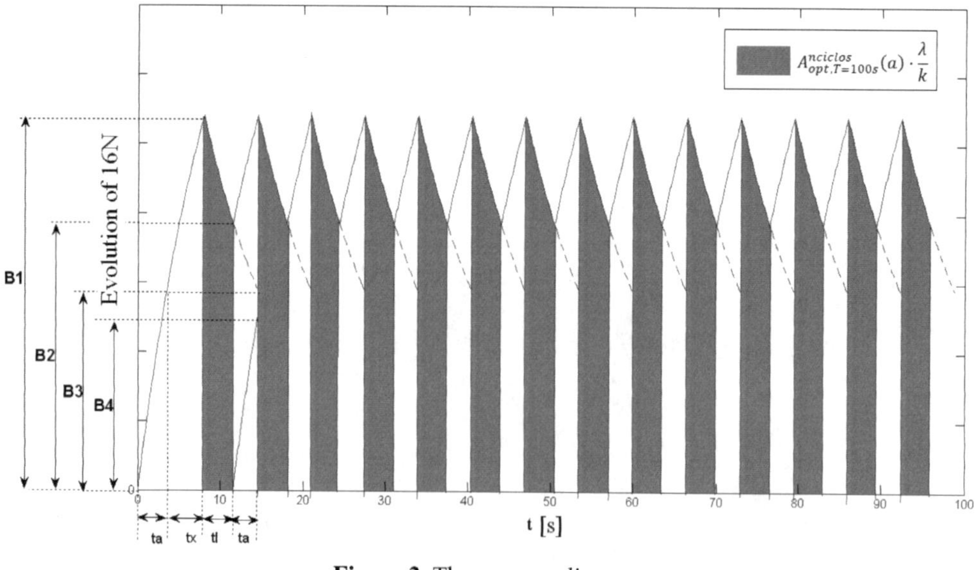

Figure 2. The *asym cyclic process*

The area in the asym cyclic activation process is given by the expression:

$$A_T^{*n\ cycles}(a) = n \cdot \left(1 - e^{-\lambda \cdot (t_a + t_x)}\right) \cdot [1 - e^{-\lambda \cdot t_l}]$$

where: $T = n \cdot (t_l + t_a) + t_x$

The comparison between the two processes is carried out under their respective optimal conditions for each one. In order to determine exactly what these optimum conditions are, it was necessary to first determine the number of cycles, activation and counting times, and the extra time for the first asym cycle that, for a determined experimental time T, will maximize the expression that defines the area mentioned above. Mathematical methods were used, the results of which are shown in Table 2. The areas for each type of activation have been called $A^{*n\ cycles}_{T,max}(s)$ and $A^{*n\ cycles}_{T,max}(a)$, and for each experimental time we have indicated the theoretical value of the area, the times, and the number of cycles at which the area is maximized. In the last column we can see the improvement that asym cyclic activation achieves over sym cyclic activation.

T [s]	*sym cycle*			*asym cycle*					*G*
	$A^{*n\ cycles}_{T,max}(s)$	τ *[s]*	*n*	$A^{*n\ cycles}_{T,max}(a)$	t_l[s]	t_a[s]	t_x[s]	*n*	*asym/sym*
50	0.9729	12.5	2	1.0750	12.47	11.3	2.558	2	1.10
100	1.9679	12.5	4	2.2800	16.72	16.04	1.82	3	1.16
200	3.9579	12.5	8	4.6460	16.61	16.43	1.859	6	1.17
300	5.9479	12.5	12	6.9800	16.71	16.43	1.838	9	1.17
400	7.9379	13.33	15	9.6900	15.5	15.12	2.039	13	1.22
500	9.9279	13.16	19	12.6100	16.61	16.6	1.866	15	1.27
600	11.9179	13.04	23	16.0100	15.16	14.74	2.098	20	1.34

Table 2. Theoretical gain of *asym* cyclic activation over *sym* cyclic activation

The theoretical gain values for each experimental time T are graphed in Figure 3, where it can be seen how they increase with time T.

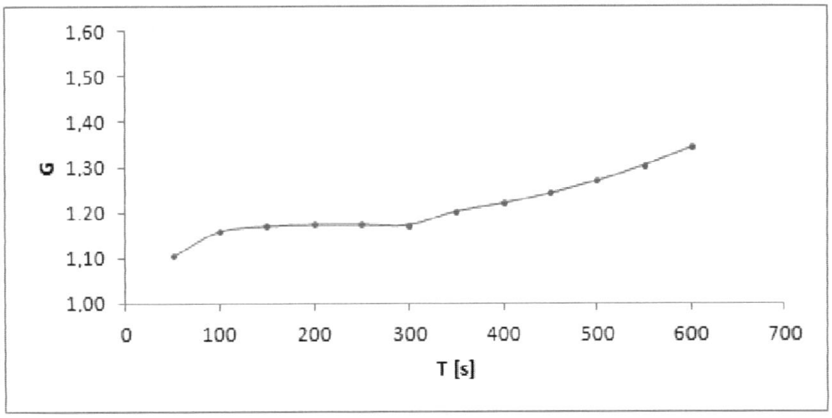

Figure 3. Graph showing the theoretical gain of *asym* cyclic activation over *sym* cyclic activation.

3. Conclusions

In conclusion, cyclic neutron activation provides a significant improvement over single-cycle neutron activation. Within the cyclic category, asym cyclic activation produces a greater concentration of 16N than sym cyclic activation, meaning that the fluorite grade in the sample can be better determined, as the number of counts increases.

4. Acknowledgments

This research was supported and financed by European Regional Development Fund (FEDER 2007-2013) from the Government of the Principality of Asturias (PCTI).

The authors would like to thank the company MINERSA SA, the Government of the Principality of Asturias, and University of Oviedo for their collaboration and financial support throughout the development of this study.

References

[1] Rey-Ronco M.A., Alonso-Sánchez T., Castro-García M.P. Neutron activation of deferred gamma rays: steps to determine the usefulness of this technique in analyzing elements of a mineral sample. Journal of Physics: Conference Series **325** 012025 (2011).

[2] Rey-Ronco M.A., Alonso-Sánchez T., Castro-García M.P. A neutron activation technique based on deferred gamma rays applied to the quantification of fluorine in fluorspar mining samples. Journal of Physics: Conference Series **325** 012021 (2011).

[3] Alonso-Sánchez T., Rey-Ronco M.A., Castro-García M.P. A neutron activation technique for the analysis for fluorine in fluorspar samples. International Journal of Mineral Processing **94** (2010) 1-13.

[4] Rey-Ronco M.A., Alonso-Sánchez T., Castro-García M.P. Production of 16N and obtaining of its gamma spectrum in order to calibrate detectors or determination of fluorine in geological specimens. Nuclear Instruments and Methods in Physics Research Section B: Beam Interactions with Materials and Atoms **268** (2010) 2766-2772.

[5] Rey-Ronco M.A., Alonso-Sánchez T., Castro-García M.P. Modeling of neutron activation process with Americium Beryllium source. Application to the activation of fluorspar samples. Journal of Mathematical Chemistry **50** 2 (2010) 325-334.

[6] Rey-Ronco M.A., Alonso-Sánchez T., Castro-García M. P. Mathematical study to improve the sensitivity in the neutron activation analysis of fluorspar. Journal Mathematical Chemistry **48** (2010) 165-174.

[7] Rey-Ronco M.A., Alonso-Sánchez T., Castro-García M.P. Modeling of neutron activation process with Americium Beryllium source. Application to the activation of fluorspar samples. Journal of Mathematical Chemistry **50** 2 (2010) 3

Application of XLENS® program to neutron diffraction data: solving crystal structures with positive and negative scatterers

J. Rius, C Frontera*, C. Miravitlles
Institut de Ciència de Materials de Barcelona-CSIC, Campus Universitari de Bellaterra, 08193 Bellaterra, Catalonia, Spain

Abstract. Direct methods based on the origin-free modulus sum function (*S*) were first adapted in 1994 to the processing of intensity data from density functions with positive and negative scatterers. Nevertheless, that implementation used phase relationships explicitly with the inherent limitation introduced by the time consuming manipulation of quartets. This limitation was removed with the introduction of the *S*-FFT algorithm (maximising *S* with only Fourier transforms) and its posterior adaptation S_2-FFT for non-positive definite density functions. In the present work both algorithms are reformulated into a generalized one, thus simplifying their implementation in the XLENS® program. The resulting unified *S*-FFT algorithm is highly effective for crystal structures with at least one medium-heavy scatterer in the unit cell. We have successfully applied it on neutron diffraction data of compounds with negative Fermi lengths.

1. Introduction

XLENS® is a computer program intended to solve crystal structure from diffraction data by applying first principles (direct methods) to the measured intensities. This is achieved by optimizing the origin-free modulus sum function, $S(\Phi)$, in terms of the Φ subset of $\varphi_{\mathbf{h}}$ phases of the **h** strong reflections [1]. Initially, $S(\Phi)$ was expressed as a function of triple-phase relationships (triplets) and hence could only handle positive definite density functions. The use of triplets made the phase refinements time-consuming. Even more computing-time demanding was the adaptation of $S(\Phi)$ to density functions with positive and negative scatterers, since in this latter case four-phase relationships (quartets) were required [2]. To overcome this limitation, the algorithm was modified to replace the need of triplets/quartets by the use of the Fast Fourier Transform (FFT) subroutine. The resulting *S*-FFT algorithm was developed in Ref [3] and extended later (S_2-FFT) to density functions with positive and negative scatterers in Refs. [4,5]. At present these are the algorithms implemented in XLENS®.

In the present contribution the description of the S-FFT and S_2-FFT algorithms is unified by making use of the recently introduced $\delta(\boldsymbol{r})$ function [6]

$$\delta(\boldsymbol{r}) = \frac{1}{V}\sum_{\mathbf{H}}(E_{\mathbf{H}} - \langle E \rangle) \cdot e^{i\alpha_{\mathbf{H}}} \cdot e^{-i2\pi\mathbf{H}\boldsymbol{r}} \quad (1)$$

which has been demonstrated to be essentially equivalent to the density function

$$\rho(\boldsymbol{r}) = \frac{1}{V}\sum_{\mathbf{h}} E_{\mathbf{h}} \cdot e^{i\varphi_{\mathbf{h}}} \cdot e^{-i2\pi\mathbf{h}\mathbf{r}} \quad (2)$$

where $E_{\mathbf{h}}$ and $\varphi_{\mathbf{h}}$ are, respectively the amplitude and the associated phase of the quasi-normalized structure factor. The E values can be derived from the corresponding experimental structure factor amplitudes F by means of $E_{\mathbf{h}} = F_{\mathbf{h}}/\sum_j f_j^2$ where f_j is the atomic scattering factor for atom j in the unit cell with N atoms (for the case of neutrons, Fermi length of nuclei must be used). If $\rho(\boldsymbol{r})$ is positive definite the relation between both is

$$\rho(\boldsymbol{r}) \cong k \cdot \delta(\boldsymbol{r}) \cdot m(\boldsymbol{r}) \quad (3)$$

with k being a suitable scaling constant and with the $m(\boldsymbol{r})$ mask taking only the values one or zero depending if the value of $\delta(\boldsymbol{r})$ is above or below the experimental threshold limit defined by $t{\cdot}\sigma(\delta)$ with t= 2.5. The $\alpha_{\mathbf{H}}$ phase values in (1) are estimated from the N highest peaks found in $\delta(\boldsymbol{r}) \cdot m(\boldsymbol{r})$. The new structure factor estimate $\boldsymbol{E}_{\mathbf{H}}$ is then given by

$$\boldsymbol{E}_{\mathbf{H},\text{new}} = E_{\mathbf{H}} \cdot e^{i\varphi_{\mathbf{H}}} = \frac{k}{V}\int_V \delta(\boldsymbol{r}) \cdot m(\boldsymbol{r}) \cdot e^{i2\pi\mathbf{Hr}} d\boldsymbol{r} \quad (4)$$

The recursive application of (4) is the basis of the δ recycling phase-refinement procedure [7].

2. The S-FFT algorithm for positive-definite ρ functions:

Let the $\rho(\boldsymbol{r})$ density function be peaked at the atomic positions. This means that ρ and the corresponding squared density function ρ^{sq} are proportional. By multiplying both sides of (3) with $\rho(\boldsymbol{r})$, it follows

$$\rho^{sq}(\boldsymbol{r}) \cong k \cdot \rho(\boldsymbol{r}) \cdot \delta(\boldsymbol{r}) \cdot m(\boldsymbol{r}) \quad (5)$$

The structure factors of ρ^{sq} may be approximated by

$$\boldsymbol{E}^{sq}_{\mathbf{h},\text{new}} = \frac{k}{V}\int_V \delta(\boldsymbol{r}) \cdot \rho(\boldsymbol{r}) \cdot e^{i2\pi\mathbf{hr}} d\boldsymbol{r} \quad (6)$$

because the mask can be set equal to 1 in the whole unit cell due to the positivity. In view of the proportionality of ρ^{sq} and ρ, the new phase estimates of ρ can be obtained from

$$\boldsymbol{E}_{\mathbf{h},\text{new}} = E_{\mathbf{h},\text{new}} \cdot e^{i\varphi_{\mathbf{h},\text{new}}} \propto \boldsymbol{E}^{sq}_{\mathbf{h},\text{new}} \quad (7)$$

which is only valid for the subset of large **h** reflections. The $\alpha_{\mathbf{H}}$ phase values associated with $\delta(\boldsymbol{r})$ are obtained from $\rho(\boldsymbol{r})^2$, i.e. from squaring the values of the $\rho(\boldsymbol{r})$ synthesis computed with the **h** reflections, only. Since no atomic peaks need be identified, no peak search subroutine is necessary.

3. The S-FFT algorithm for sign-unrestricted ρ functions

In the case of ρ functions with positive and negative peaks, it is convenient to introduce a new mask fulfilling the following criterion [4,5]:

$$m(\mathbf{r}) = \begin{cases} 1 & \text{if } \rho(\mathbf{r}) > t\,\sigma(\rho) \\ -1 & \text{if } \rho(\mathbf{r}) < -t\,\sigma(\rho) \\ 0 & \text{if } -t\,\sigma(\rho) < \rho(\mathbf{r}) < t\,\sigma(\rho) \end{cases} \quad (8)$$

where **r** ranges over the whole unit cell and with t being close to 2.5. This mask allows introducing a new density function called $\rho^{sq-s}(\boldsymbol{r})$ where *sq-s* stand for "squared-shaped". This new function results from the product of $\rho^{sq}(\boldsymbol{r}) \cdot m(\boldsymbol{r})$ and has the same peak shapes than $\rho^{sq}(\boldsymbol{r})$ but the same peak signs as $\rho(\boldsymbol{r})$. In other words, $\rho^{sq-s}(\boldsymbol{r})$ and $\rho(\boldsymbol{r})$ are proportional. By multiplying both sides of (3) with $\rho(\boldsymbol{r}) \cdot m(\boldsymbol{r})$, it follows

$$\rho^{sq-s}(\boldsymbol{r}) \cong k \cdot \rho(\boldsymbol{r}) \cdot \delta(\boldsymbol{r}) \cdot m(\boldsymbol{r}) \quad (9)$$

with the structure factors of $\rho^{sq-s}(\boldsymbol{r})$ being

$$\boldsymbol{E}^{sq-s}_{\mathbf{h},new} = \frac{1}{V}\int_V \delta(\boldsymbol{r}) \cdot \rho(\boldsymbol{r}) \cdot m(\boldsymbol{r}) \cdot e^{i2\pi\mathbf{hr}} d\boldsymbol{r} \quad (10)$$

Due to the proportionality of ρ^{sq-s} and ρ, the new phase estimates of ρ can be obtained from

$$\boldsymbol{E}_{\mathbf{h},\text{new}} = E_{\mathbf{h},\text{new}} \cdot e^{i\varphi_{\mathbf{h},\text{new}}} \propto \boldsymbol{E}^{sq-s}_{\mathbf{h},\text{new}} \quad (11)$$

which only holds for the subset of large **h** reflections. The α_H phase values associated with $\delta(\boldsymbol{r})$ are obtained by squaring the values of the $\rho(\boldsymbol{r})$ synthesis computed with only the **h** reflections followed by multiplication with $m(\mathbf{r})$.

4. The unified *S*-FFT algorithm

Figure 1 shows the flow chart of the S-FFT algorithm in its unified version. Definition of the mask **m** is in accordance to the type of problem being treated: It takes values 1 or 0 (for ρ above or below the threshold limit, respectively) in the case of positive scatterers, or it is defined by (8) in the case of positive and negative scatterers. A slight modification of *m* was shown to be more effective in the case of light scatterers [5]

$$m(\mathbf{r}) = 1 \quad \text{if } \rho(\mathbf{r}) > t\,\sigma(\rho) \qquad (12)$$
$$= -1 \quad \text{if } \rho(\mathbf{r}) < -t\,\sigma(\rho)$$
$$= a \quad \text{if } -t\,\sigma(\rho) < \rho(\mathbf{r}) < t\,\sigma(\rho)$$

where *a* is a random number between -1 and 1. Nevertheless, this modification was tested for solving organic compounds from X-ray diffraction data but not for solving structures from neutron data. In the following section we have tested this new mask for two examples from neutron diffraction.

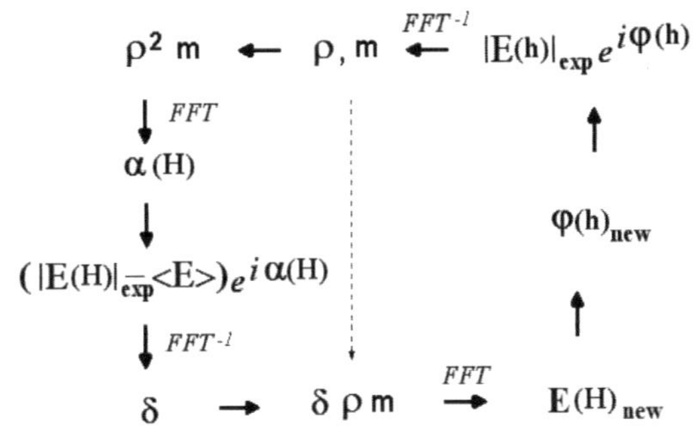

Figure 1

The unified *S*-FFT phase refinement procedure. The initial phases (upper right corner) are combined with the experimental amplitudes to compute the electron density ρ and the associate mask m. The mask is used to compute the phases α of the δ function and also (broken arrow) the product function, $\delta{\cdot}\rho{\cdot}m$. The Fourier transform of this product function yields the new structure factor estimates.

5. Results of test structure solutions from neutron diffraction.

5.1. $Bi_{0.75}Sr_{0.25}MnO_3$

We have used S-FFT implemented in XLENS® (option MASK=-2) to find negative neutron scatterers in two different examples of well known structures. The first example consists on its application to a perovskite compound with unit formula $(Bi_{0.75}Sr_{0.25})MnO_3$ containing the strong negative neutron scatterer Mn [8]. Fermi lengths for Bi, Sr, Mn and O are, respectively, 0.853, 0.702, -0.373, 0.580 fm and the crystal data are *a*= 5.499, *b*= 7.770, *c*= 5.542 Å, space group: *I mma*, *Z*= 4. We have used in the calculations the intensities extracted from the observed powder diffraction pattern (measured at D2B) by redistributing the global intensities of the overlapped peaks according to the calculated individual intensities. After 50 trials (starting with random phases), the program reports the true structure for 7 sets of initial values. These right solutions are easily identifiable for their low values of the residual. Table 1 lists the peaks in the E-map. Bond distances found directly from the structural resolution are listed and compared with those reported in Ref. [8] as a test for the accuracy of the peak finding.

Table 1. Application of the unified *S*-FFT algorithm to neutron diffraction data. Compound data: a=5.499, b= 7.770, c=5.542 space group: *I mma*, Unit formula: $(Bi_{0.75}Sr_{0.25})MnO_3$, Z=4. Success rate 3/25. All atoms with absolute relative height greater than 90 are listed. Bond distances found are compared with those reported from the refinement of NPD data [8].

Atom	rel. height	x/a	y/b	z/c	Site	Bond distances(Å)	
Bi,Sr	1000	0	¼	0.9974	4e	Found/Reported	
Mn	-579	½	0	0	4b	Mn-O1	⟨Mn-O2⟩
O1	474	½	¾	0.0501	4e	1.962	1.964
O2	409	¼	0.0275	¼	8g	1.9690(9)	1.9704(6)

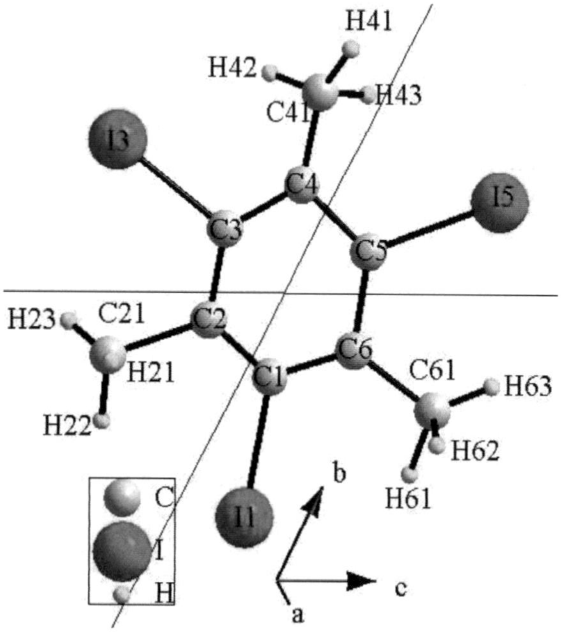

Figure 2:
Graphical view of the C9H9I3 compound from the peak positions found by application of extended S_2-FFT procedure implemented in XLENS®.

Table 2
C-H bond distances (in Å) as found compared to that refined in Ref. [9].

Bond	Found	Report.
C21-H21	0.952	1.068(4)
C21-H22	1.176	1.076(4)
C21-H23	0.853	1.068(4)
C41-H41	1.066	1.077(3)
C41-H42	0.911	1.059(5)
C41-H43	1.057	1.046(5)
C61-H61	1.173	1.081(4)
C61-H62	0.973	1.085(4)
C61-H63	0.958	1.044(4)

5.2. $C_9H_9I_3$

The second test example is the application of XLENS® to the single-crystal neutron diffraction data of the 1,3,5-triiodo-2,4,6-trimethylbenzene (TIM) molecule. The single-crystal neutron diffraction study of this molecule was published by [9]. The neutron intensities were collected at 15 K on the D9 four-circle diffractometer at the high-flux reactor of ILL (Grenoble, France). The monochromator was a copper single-crystal selecting the (2 2 0) reflection giving λ= 0.70379 Å. Crystal data: a = 7.905, b = 9.510, c = 9.521 Å, α= 60.41, β= 66.64, γ= 86.24°, V= 563.7A3, space group P-1, molecular formula: $C_9H_9I_3$, Z= 2, crystal dimensions: 4 x 2 x 1 mm^3. Due to detector problems only 2071 reflections could be measured (-4< h <8, -9< k <9, -9< l <9). Averaging of the equivalent data gave 1079 independent reflections. The Fermi lengths for C, I and H are, respectively, 0.665, 0.528 and -0.374. The parameters defining the direct methods *S* phase refinement function are: <E> = 1.073, d_{min}=0.82, E_{ratio}=0.92, strongest E value = 3.46, lowest large E value= 1.35, number of sets: 100, maximum number of cycles: 39. Phase refinement with *S*-FFT in *P*1 always gives the correct solution. The subsequent automated finding of the symmetry centre is successful in 30% of the cases. The structure found is depicted in Fig. 2 and it is in very good accordance with the refined structure (Fig. 7 in Ref. [9]). Bond distances found directly from the structural resolution are listed in Table 2 and compared with those reported from the structural refinement.

6. Conclusions

The introduction of mask m(**r**) (8) or (12) in the procedure depicted in Fig. 1 allows the proper finding of negative scatterers. To prove it and show its applicability to the solution of structures containing nuclei with negative Fermi lengths, we have used XLENS® in two case examples and successfully found Mn (in the first) and H (in the second). The comparison of bond distances with those reported prove that the positions found for negative scatterers are accurate enough and can be taken as a good starting point for subsequent structural refinement. This proves the utility of this implementation for solving structures from neutron diffraction.

Acknowledgments

The authors thank the Spanish 'Ministerio de Ciencia e Innovación' (projects MAT2009-07967 and Consolider NANOSELECT CSD2007-00041) and the 'Generalitat de Catalunya' SQR(2009) for financial support. The authors want to express their gratitude to Dr. Mª Teresa Fernández-Díaz for kindly supplying the single-crystal neutron diffraction data.

References

[1] Rius J, 1993, Acta Cryst. A**49**, 406.
[2] Rius J, Miravitlles C, Allman R, 1996, Acta Cryst. A**52**, 634.
[3] Rius J, Crespi A, Torrelles X, 2007, Acta Cryst. A**63**, 131.
[4] Rius J, Frontera C, 2008, Acta Cryst. A**64**, 670.
[5] Rius J, Frontera C, 2009, Acta Cryst. A**65**, 528.
[6] Rius J, 2012, Acta Cryst. A**68**, 77.
[7] Rius J, 2012, Acta Cryst. A**68**, 399.
[8] Frontera C, García-Muñoz J L, Aranda M A G, Hervieu M, Ritter C, Mañosa L, Capdevila X G, Calleja A, 2004, Phys. Rev. B **68**, 134408
[9] Boudjada A, Meinnel J J, Boucekkine A, Hernandez O J, Fernández-Díaz M T, 2002, J. Chem. Phys. **117**, 10173.

Neutron Spectroscopy as a Probe of Macromolecular Structure and Dynamics under Extreme Spatial Confinement

F Barroso-Bujans,[1,*] F Fernandez-Alonso[2,3] and J Colmenero[1,4,5]

[1]Centro de Física de Materiales (CSIC-UPV/EHU)-Material Physics Center (MPC), Paseo Manuel Lardizábal 5, 20018 San Sebastian, Spain

[2]ISIS Facility, Rutherford Appleton Laboratory, Chilton, Didcot, Oxfordshire OX11 0QX, United Kingdom

[3]Department of Physics and Astronomy, University College London, Gower Street, London, WC1E 6BT, United Kingdom.

[4]Departamento de Física de Materiales, Universidad del País Vasco (UPV/EHU) Aptdo 1072, 20080 San Sebastian, Spain

[5]Donostia International Physics Center, Paseo Manuel Lardizábal 4, 20018 San Sebastian, Spain

E-mail: `fbarroso@ehu.es`

Abstract. We illustrate the use of high-resolution neutron spectroscopy to explore the extreme spatial confinement of soft matter in nanostructured materials. Two well-defined limits are considered, involving either intercalation or interfacial adsorption of the ubiquitous polymer poly(ethylene oxide) in graphite-oxide-based hosts. Vibrational modes associated with the confined macromolecular phase undergo dramatic changes over a broad range of energy transfers, from those associated with intermolecular modes in the Terahertz frequency range (1 THz = 33 cm^{-1}), to those characteristic of strong chemical bonds above 2000 cm^{-1}. We also consider the effects of polymer chain size and chemical composition of the host material. Variation of the degree of oxidation and exfoliation of graphite oxide leads to two distinct cases, namely: (i) subnanometer two-dimensional confinement; and (ii) surface immobilization. Case (i) is characterised by significant changes to conformational and collective vibrational modes of the polymer as a consequence of a preferentially planar trans-trans-trans chain conformation, whereas case (ii) leads to a substantial increase in the population of gauche conformers. Macroscopically, case (i) translates into the complete suppression of crystallization and glassy behaviour. In contrast, case (ii) exhibits well-defined glass and melting transitions associated with the confined phase, yet at significantly lower temperatures than those of the bulk.

1. Introduction

Optical spectroscopy has been used extensively to characterise polymeric and macromolecular materials [1]. Traditionally, Fourier-transform Infrared (FTIR) and Raman spectroscopy have been the most versatile methods to identify specific structural features related to polymer conformation in the solid and molten phases. The ubiquitous polymer poly(ethylene oxide), hereafter PEO, is a good case in point, as the solid phase is characterised by a distinct helical structure which is only retained partially in the liquid or under conditions of strong spatial confinement [2, 3, 4].

Notwithstanding the above, optical spectroscopic techniques do fail in some important cases of increasing scientific and technological interest. Such is the case of macromolecular intercalation or adsorption in layered and nanostructured host media, particularly when the latter is characterised by a strong optical response in the spectral range of interest. Graphite-based substrates fall into this problematic class of materials owing to their strong absorption in the infrared [5] or electronic (resonant) enhancement effects in the visible region of the electromagnetic spectrum [6, 7]. Owing to the strong neutron-proton incoherent cross section, inelastic neutron scattering (INS) can circumvent these difficulties quite elegantly in the microscopic characterisation of confined hydrogenous matter. We have recently demonstrated how broadband, high-resolution INS, as that afforded by the TOSCA spectrometer at the ISIS Facility (United Kingdom) [8], constitutes an exquisite characterisation tool to study soft hydrogenous matter confined in graphite-oxide (GO) materials [9, 10, 11].

In this contribution to the Proceedings of the VI Meeting of the Spanish Neutron Scattering Association, we highlight the merits and strengths of INS as a tool to characterise molecular and macromolecular structure and dynamics under extreme spatial confinement. To this end, we compare INS data obtained for PEO chains of increasing length, from a single monomeric unit [i.e., ethylene glycol (EG)] to several thousand repetitive ethylene-oxide units. Furthermore, by varying the degree of GO oxidation and exfoliation, we also explore distinct regimes of spatial confinement, from subnanometer two-dimensional confinement to surface immobilization onto graphene (G) sheets. We close by outlining both the basic instrumental prerequisites that have enabled these studies as well as some ways forward.

2. Experimental

2.1. Synthesis of graphite-oxide-based materials and intercalation compounds

GOs were produced by oxidation of natural graphite using a modified Brodie method [12]. Varying degrees of GO oxidation were obtained by changing the reaction time and temperature [9, 13]. Specimen G was obtained by placing GO on a glass boat and inserting it in a quartz tube under an argon flux. This tube was then inserted into a tube furnace preheated to 1000 °C. After 1 min, the tube was removed and cooled down to room temperature. Pentaethylene glycol (5PEO) was purchased from Fluka, whereas EG and PEO [$HO(CH_2CH_2O)_nH$] with chain lengths n = 13, 104, 795, and 2135 were supplied by Aldrich. Hereafter, we will use the term "nPEO" to identify a PEO chain of length "n." Intercalation of EG and 5PEO was performed by direct mixing with GO in the absence of solvent, whereas intercalation of 13PEO and 2135PEO was carried out in aqueous solution by stirring a total of 0.5 g PEO previously dissolved in 20 mL water with 0.5 g GO over a period of 15 days. A similar procedure was followed to absorb 2135PEO onto the G specimen. Any EG and PEO excess was removed by filtration and thorough aqueous washings. The resulting EG/GO, nPEO/GO (n=5, 13 and 2135) and 2135PEO/G specimens were then dried at 80°C for 24 h in a vacuum oven (P < 0.1 mbar). The amount of EG(nPEO) in EG(nPEO)/GO and nPEO/G was calculated from thermogravimetric sample-residue analysis as described previously in Refs. [10, 14]. Additional compositional and structural information can be found in Ref. [13].

2.2. Characterisation

INS experiments were performed on the TOSCA spectrometer [8] located at the ISIS Facility, Rutherford Appleton Laboratory, United Kingdom. Neutron time-of-flight data were collected in both back- and forward-scattering geometries, and then added together to obtain a hydrogen-projected vibrational density of states (VDOS). Typical run times varied between 2 and 8 h depending on the hydrogen content of the sample. All samples were contained in flat aluminium

Journal of Physics: Conference Series **549** (2014) 012009 doi:10.1088/1742-6596/549/1/012009

cells of thickness 1-4 mm and cooled to temperatures below 30 K. Based on the thermal behaviour of EG and nPEO observed by differential scanning calorimetry (DSC), EG and nPEO with n $\geq$ 13 cannot be obtained in an amorphous state, whereas the quenching of molten 5PEO in liquid nitrogen can prevent crystallisation. Amorphous 5PEO was then produced by direct immersion (quenching) into the TOSCA cryostat (T $<$ 30K). Crash cooling to cryogenic temperatures (T $<$ 100K) was optimised by placing the samples at the bottom of the cryostat and exposing them directly to a stream of liquid helium. To prepare a crystalline sample of 5PEO, the specimen was first crystallised in a freezer held at 240K, followed by immersion in the low-temperature cryostat. INS data of GO, G, EG, and nPEO were normalized to sample mass and those of EG(nPEO)/GO and 2135PEO/G were normalized to the amount of EG(nPEO) content.

3. Results and Discussion

3.1. Molecular-size effects

Both EG and nPEO intercalate into the GO interlayer space giving rise to a single molecular layer of thickness 3.0-3.4 Å. Their concentration in the intercalate amounts to 22$\pm$2 wt% of the total EG(nPEO)/GO mass. Both the interlayer spacing and the amount of EG and nPEO do not depend strongly on intercalate chain length, a strong indication that the oligomer and polymer chains are forced to adopt a planar conformation in a monolayer arrangement, i.e., well-defined and extreme two-dimensional (2D) confinement. Moreover, neither EG nor nPEO can crystallize in the intercalate and their corresponding glass transitions are not present in the calorimetric data. Previous dielectric-spectroscopy studies to probe the α-relaxation process (associated with the glass transition) and other processes related to molecular mobility below the glass transition have also shown the suppression of α-relaxation phenomena as well as a slowdown of β-relaxation modes for 2135PEO [11]. These results were rationalized in terms of a complete suppression of cooperativity among polymer chains and restricted chain motions induced by strong (hydrogen-bonding) PEO-GO interactions. On the basis of the above experimental results, the cartoon in Figure 1 provides a schematic representation of the EG/GO and 5PEO/GO materials of relevance to the present study.

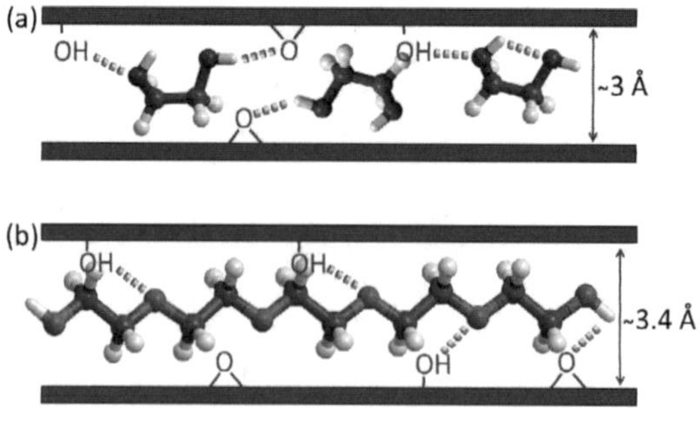

Figure 1. Sketch of (a) EG/GO and (b) 5PEO/GO intercalates. Dashed lines show the intermolecular hydrogen bonds between the EG(5PEO) intercalate and the GO host.

Figure 2 shows the attenuated-total-reflectance Fourier-transform infrared (ATR-FTIR) spectrum of GO, as well as those of bulk and confined PEO with the longest chain length of the series (2135PEO). As observed, FTIR spectral features associated with confined PEO are severely affected by the presence of a highly absorbing GO matrix, making it practically impossible to extract quantitative information on changes to the vibrational spectrum of the polymer upon intercalation. The most significant changes in these ATR-FTIR data take place in the OH stretch region at ca. 3250 cm^{-1}, associated with the presence of hydroxyl (OH) groups in the GO substrate. The red shift observed in going from pristine GO to 2135PEO/GO indicates the formation of hydrogen bonds between OH groups in GO and PEO ether groups, and are associated with an overall weakening of O-H bonds [15]. This assignment is also supported by a red shift of the C-O-C stretching vibration band centred at 1103 cm^{-1} in bulk PEO. The FTIR data thus confirms the presence of specific intermolecular interactions between PEO and GO, yet it provides no direct information about the confined polymer phase.

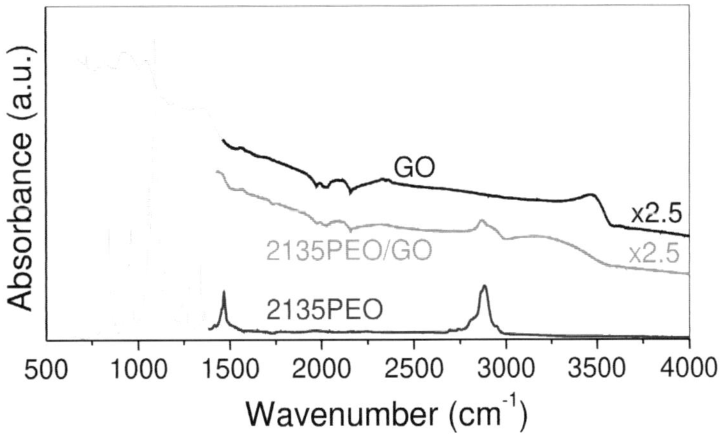

Figure 2. Room-temperature ATR-FTIR spectra of GO, bulk 2135PEO, and 2135PEO/GO. For ease of comparison, the overall intensities of the spectra corresponding to GO and 2135PEO/GO have been multiplied by a factor of 2.5.

To gain further insight into the macromolecular structure of confined EG and nPEO, INS measurements on EG/GO, nPEO/GO, and their corresponding pristine materials were performed on the TOSCA spectrometer at 30 K. One of the primary advantages of INS spectroscopy is that there are no hard selection rules and mode intensities can be directly related to the underlying VDOS. Therefore, spectral assignments can be performed on the basis of our current understanding of the vibrational spectrum of the bulk polymer obtained from optical spectroscopy [16, 17, 18] and computer simulation [19]. Another advantage of INS spectroscopy lies in its high sensitivity to vibrational modes involving hydrogen, thereby greatly simplifying spectral assignment when the host matrix is non hydrogenous. A detailed assignment of these bands has been reported in Refs. [10, 11]. Last but not least, INS spectroscopy allows access to very low energy transfers below ~200 cm^{-1}, difficult to access with conventional infrared and Raman techniques.

INS spectra for bulk and confined GO, EG, and nPEO (n=5,13, and 2135) over the energy-transfer range 24-4000 cm^{-1} are shown in Figure 3. The GO spectrum exhibits a very weak and featureless INS response, which enables a clear identification of EG(nPEO) bands in the EG(nPEO)/GO samples. At higher energy transfers (shaded region I in Figure 3), the INS spectra is dominated by a broad band associated with C-H stretch vibrations centred at 2930-2960 cm^{-1}. The integrated intensity of this high-energy feature corresponding to C-H stretch modes serves to validate the amount of EG(nPEO) in the sample (ca. 22 wt%), as detailed previously in reference [11] for the case of the 2135PEO/GO intercalate. Moreover, we find that this particular spectral feature does not undergo any changes in intensity or line shape upon confinement and can, therefore, be used as a reliable estimator of the total concentration of hydrogen in a given sample. On the basis of these observations, intensity changes at lower energy transfers can then be linked to changes in the vibrational dynamics of the intercalate upon confinement.

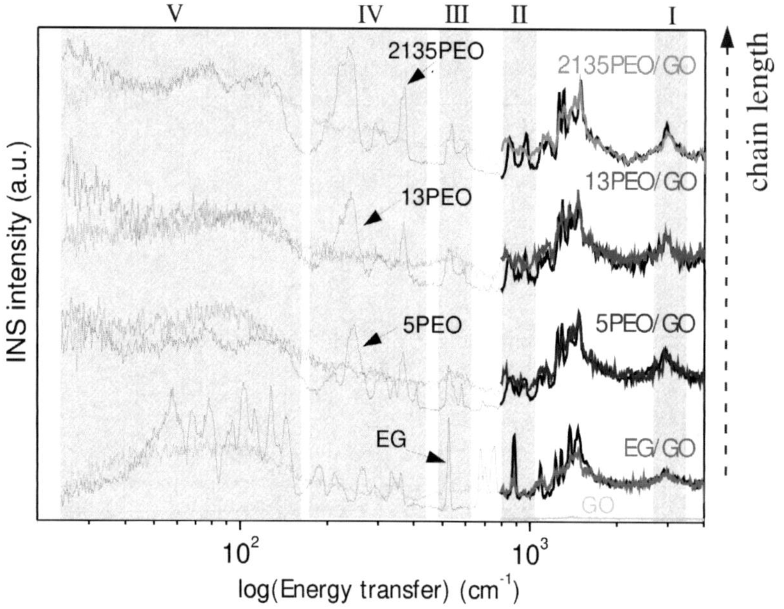

Figure 3. Mass-normalised INS spectra of GO, EG, and PEO (n = 5-2135) in the bulk and under confinement in GO (EG/GO and nPEO/GO). For an explanation of each shaded region in the spectra, see the main text.

The most significant differences between bulk and confined EG(nPEO) are observed at energy transfers below 1000 cm^{-1} (shaded regions II-V in Figure 3). INS spectral bands either disappear, shift, or broaden in confined EG(nPEO) compared to the bulk. We can rationalize these results in terms of restricted EG(nPEO) motions within the 2D GO interlayer. These spectral regions are discussed in more detail below.

Region II: CH_2 rocking modes [r(CH_2)] in the spectral range 800-1000 cm^{-1} are particularly sensitive to macromolecular conformation for nPEO (i.e., trans and gauche conformers). These spectral features are also quite sensitive to temperature, dilution, metal coordination, and geometric constraints [20, 16, 21, 17, 18]. The band at 846 cm^{-1} corresponds to *trans-gauche-trans* (*tgt*) conformations of CCOC, OCCO, and COCC groups in crystalline nPEO. This band moves down to 831 cm^{-1} in the amorphous phase for the quenchable 5PEO bulk material [10], indicating a sensible increase in the population of *trans* conformers. In intercalated nPEO, this band moves further down to 814 cm^{-1} as a result of the predominance of *trans-trans-trans* (*ttt*) geometries [18, 20]. This assignment is further corroborated by the red shift and intensity suppression of the peak at 953 cm^{-1} and 943 cm^{-1}, corresponding to *tgt* conformations in the crystal and in the amorphous phase, respectively. Table 1 summarizes the position of r(CH_2) bands for crystalline, amorphous, and confined 5PEO.

Region III: this spectral range (480-600 cm^{-1}) provides information on the molecular conformation of EG. On the basis of infrared and Raman data for liquid and solid EG [22, 23], as well as for EG confined in silica-glass pores of radii 3-7 nm [24], the INS band at 523

cm^{-1} corresponds to the asymmetric CCO bending mode of the OCCO skeleton of the gauche conformer for crystalline EG [10]. The presence of a broad INS peak at 521 cm^{-1}, as well as the absence of a peak at 482 cm^{-1}(trans conformers [22]) for EG/GO indicates that EG adopts a disordered gauche conformation upon confinement.

Table 1. INS bands corresponding to CH_2 rocking vibrations in crystalline, amorphous, and confined 5PEO.

phase	Band 1	Band 2	conformation
crystalline	953	846	*tgt*
amorphous	943	831	intermediate
confined	928	816	*ttt*

Region IV: changes in the longitudinal-acoustic-mode (LAM) region (150-400 cm^{-1}) [19] provide more evident signatures of changes to macromolecular conformation. These bands correspond to complex modes associated with collective vibrations along the polymer chain. From the INS data, the absence of peaks in the LAM region for all confined nPEO chains as compared to their bulk counterparts is quite apparent. Such a result is largely insensitive to molecular size and suggests the emergence of a new set of chain length scales primarily dictated by the presence of anchoring points on the GO substrate upon intercalation.

Region V: the low-energy bands at 74 (VI) and 115 cm^{-1} (VII) arise from either torsional COC or CO internal rotations in nPEO. These features are considerably broader in the confined phase and closely resemble the vibrational density of states of a disordered amorphous polymer [25].

3.2. Host effects

The previous section has shown that the extreme two-dimensional confinement of EG and nPEO in GO at subnanometer length scales induces profound changes to the structure and dynamics of the polymer phase. In the present section, we address the question as to whether polymer-substrate interactions or purely geometric restrictions are the driving force governing this behaviour. To this end, we prepared and investigated a series of 2135PEO/GO intercalates with a variable degree of GO oxidation, as well as a specimen consisting of 2135PEO adsorbed on thermally reduced G sheets. The amount of intercalated PEO in GO varies from 9 to 27 wt% as the degree of oxidation is increased from an oxygen-to-carbon atomic ratio (O/C) of 0.29 to 0.39 [13]. From this point onwards, the amount of intercalated PEO undergoes a slight decrease to 23 wt% at 0.40 O/C. The INS and DSC data of these materials are quite insensitive to the degree of oxidation of the underlying GO host. In contrast, these data show significant differences in the behaviour of the adsorbed 2135PEO phase on G (PEO/G) compared to the intercalated 2135PEO phase [9]. The amount of PEO adsorbed on G increases to 28 wt% in spite of the low O/C of G (0.13), a result which can be explained in terms of the high (open) surface area characteristic of this material (BET area ca. 630 cm^2/g). Moreover, the DSC data for 2135PEO exhibits a clear glass transition (T_g) at 209 K and melting (T_m) at 304 K, noticeably lower than those of the bulk material (218 and 333 K, respectively). The INS data show clear differences in the spectrum of 2135PEO/G compared to that of 2135PEO/GO (see reference [9]). In particular, the CH_2 rocking band at 815 cm^{-1} in 2135PEO/GO shifts to 835 cm^{-1}, an intermediate energy between that of the confined and the crystalline 2135PEO phase (846 cm^{-1}). This spectral shift indicates a sensible increase in the population of gauche conformers in 2135PEO/G compared to that in 2135PEO/GO. Table 2 summarises the CH_2 rocking-band positions for bulk-crystalline and confined specimens in GO and G. These findings are consistent with our DSC experiments showing clear signs of crystal melting in PEO/G. From the present comparison between GO- and G-based materials, we conclude that the geometry of the host medium plays a critical role at dictating the ultimate properties of the confined polymer phase, particularly under conditions of extreme (subnanometer) spatial confinement.

Table 2. INS bands (in cm^{-1}) corresponding to the CH_2 rocking vibrations of methylene groups in 2135PEO in the bulk crystal, intercalated in GO, and adsorbed onto G sheets.

phase	peak 1	peak 2	conformation
crystalline	946	845	*tgt*
adsorbed on G	930	835	intermediate
intercalated in GO	914	815	*ttt*

4. Perspectives

High-resolution neutron spectroscopy provides access to structural and dynamical information on confined hydrogenous materials not accessible by any other experimental probe. As the present work highlights, basic prerequisites include access to a broad range of energy transfers (0-4000 cm^{-1}) as well as tight resolution (1-2 %) across the entire spectral range. The former is of particular relevance to assess the effects of spatial confinement on both intra and intermolecular modes, whereas the latter is of paramount importance in order to follow subtle yet at the same time significant changes to macromolecular conformation in the fingerprint region of the vibrational spectrum below 1500 cm^{-1}. At the present time, the TOSCA spectrometer at the ISIS Facility provides these unique capabilities on a routine basis, yet we note that both LAGRANGE at the ILL (France) and VISION at the SNS (USA) are now in a position to tackle these challenges with success as well. The present INS measurements have been restricted to low temperatures as our primary objective was to probe the VDOS of the confined hydrogenous phase. Notwithstanding the above, extension of these high-resolution INS studies to higher temperatures also offers the exciting prospects of studying the emergence of critical phenomena such as two-dimensional melting in these complex materials. Such a task is significantly more intricate than previous studies of atomic and molecular intercalation in pristine graphite [26, 27, 28, 29], and might well require a judicious combination of both inverted- and direct-geometry INS instrumentation in order to cover a sufficiently broad spectral range with adequate resolution.

To conclude, we note that further work is also underway to validate detailed computational models of extreme molecular and macromolecular confinement in well-defined nanostructured materials using INS, as well as to link polymer conformation and vibrational structure with the dynamical response and transport properties of these novel materials at longer time scales, accessible via quasielastic neutron scattering techniques and dielectric spectroscopy.

Acknowledgements

The authors gratefully acknowledge the continued support from the Spanish Ministry of Education (MAT2007-63681), Basque Government (IT-436-07), Gipuzkoako Foru Aldundia

(2011-CIEN-000085-01), and the UK Science and Technology Facilities Council. F.B.-B. acknowledges financial support from BERC-MPC.

References

[1] Painter P C, Coleman M and Koenig J 1982 *The Theory of Vibrational Spectroscopy and its Application to Polymeric Materials* (New York: John Wiley & Sons)
[2] Aranda P and Ruiz-Hitzky E 1992 *Chem. Mater.* **4** 1395–1403
[3] Matsuo Y, Tahara K and Sugie Y 1997 *Carbon* **35** 113–120
[4] Miwa Y, Drews A R and Schlick S 2008 *Macromolecules* **41** 4701–4708
[5] Nemanich R, Lucovsky G and Solin S 1977 *Solid State Commun.* **23** 117–120
[6] Ferrari A C 2007 *Solid State Commun.* **143** 47–57
[7] Reich S and Thomsen C 2004 *Philos. Trans. Roy. Soc. London. Ser. A Math. Phys. Eng. Sci.* **362** 2271–2288
[8] Colognesi D, Celli M, Cilloco F, Newport R J, Parker S F, Rossi-Albertini V, Sacchetti F, Tomkinson J and Zoppi M 2002 *Appl. Phys. A-Mater. Sci. Process.* **74** s64
[9] Barroso-Bujans F, Fernandez-Alonso F, Pomposo J A, Cerveny S, Alegria A and Colmenero J 2012 *ACS Macro Lett.* **1** 550–554
[10] Barroso-Bujans F, Fernandez-Alonso F, Cerveny S, Arrese-Igor S, Alegría A and Colmenero J 2012 *Macromolecules* **45** 3137–3144
[11] Barroso-Bujans F, Fernandez-Alonso F, Cerveny S, Parker S F, Alegría A and Colmenero J 2011 *Soft Matter* **7**(16) 7173–7176
[12] Brodie B C 1859 *Philos. Trans. Roy. Soc. London* **149** 249–259
[13] Barroso-Bujans F, Fernandez-Alonso F, Pomposo J A, Enciso E, G Fierro J L and Colmenero J 2012 *Carbon* **50** 5232–5241
[14] Barroso-Bujans F, Alegria A, Pomposo J A and Colmenero J 2013 *Macromolecules* **46** 1890–1898
[15] Cerveny S, Barroso-Bujans F, Alegría A and Colmenero J 2010 *J. Phys. Chem. C* **114** 2604–2612
[16] Yoshihara T, Tadokoro H and Murahashi S 1964 *J. Chem. Phys.* **41** 2902–2911
[17] Matsuura H and Miyazawa T 1969 *J. Polym. Sci., Part A2: Polym. Phys.* **7** 1735–1744
[18] Maxfield J and Shepherd I W 1975 *Polymer* **16** 505–509
[19] Yang X, Su Z, Wu D, Hsu S L and Stidham H D 1997 *Macromolecules* **30** 3796–3802
[20] Papke B L, Ratner M A and Shriver D F 1981 *J. Phys. Chem. Solids* **42** 493–500
[21] Koenig J L and Angood A C 1970 *J. Polym. Sci., Part A2: Polym. Phys.* **8** 1787–1796
[22] Matsuura H and Miyazawa T 1967 *Bull. Chem. Soc. Jpn.* **40** 85–94
[23] Matsuura H, Hiraishi M and Miyazawa T 1972 *Spectrochim. Acta, Part A* **28** 2299–2304
[24] Luo R S and Jonas J 2001 *J. Raman Spectrosc.* **32** 975–978
[25] Frick B and Fetters L J 1994 *Macromolecules* **27** 974–980
[26] Enoki T, Endo M and Suzuki M 2003 *Graphite intercalation compounds and applications* (Oxford University Press, Incorporated)
[27] Simon C, Rosenman I, Batallan F, Lartigue C and Legrand J F 1992 *Phys. Rev. B* **45**(6) 2694–2698
[28] Simon C, Rosenman I, Batallan F, Rogerie J, Legrand J F, Magerl A, Lartigue C and Fuzellier H 1990 *Phys. Rev. B* **41**(4) 2390–2397
[29] Simon C, Rosenman I, Batallan F, Pepy G and Lauter H 1988 *Synth. Met.* **23** 147–153

Journal of Physics: Conference Series **549** (2014) 012010 doi:10.1088/1742-6596/549/1/012010

Modulating Pluronics micellar rupture with cyclodextrins and drugs: effect of pH and temperature

M Valero[1] and C A Dreiss[2]

[1]Dpto. Química Física, Facultad de Farmacia, Universidad de Salamanca, Campus Miguel de Unamuno, 37008 Salamanca, Spain
[2] Institute of Pharmaceutical Science, King's College London, Franklin-Wilkins Building, 150 Stamford Street, SE1 9NH London, U.K.

Email: mvalero@usal.es

Abstract. Micelles of the triblock copolymer Pluronic F127 can encapsulate drugs with various chemical structures and their architecture has been studied by small-angle neutron scattering (SANS). Interaction with a derivative of β-cyclodextrin, namely, heptakis(2,6-di-*O*-methyl)-β-cyclodextrin (DIMEB), induces a complete break-up of the micelles, providing a mechanism for drug release. In the presence of drugs partitioned within the micelles, competitive interactions between polymer, drug and cyclodextrin lead to a modulation of the micellar rupture, depending on the nature of the drug and the exact composition of the ternary system. These interactions can be further adjusted by temperature and pH. While the most widely accepted mechanism for the interaction between Pluronics and cyclodextrins is through polypseudorotaxane (PR) formation, involving the threading of β-CD on the polymer backbone, time-resolved SANS experiments show that de-micellisation takes place in less than 100 ms, thus unambiguously ruling out an inclusion complex between the cyclodextrin and the polymer chains.

1. Introduction

Triblock copolymers of poly(ethylene oxide)-poly(propylene oxide)-poly(ethylene oxide) (PEO-PPO-PEO), known as Pluronics or Poloxamers, are polymeric surfactants able to form micellar aggregates in solution [1]. They can solubilise drugs in an aqueous environment, and present similar features to natural carriers in terms of size, structure and transport properties, making them excellent candidates for drug delivery applications [2].

Natural cyclodextrins are cyclic oligosaccharides, formed by 6, 7 and 8 glucose units: α-, β- and γ-cyclodextrins, respectively [3]. Their derivatives, obtained by modifying the external hydroxyl groups with various types of substituents (such as hydroxyethyl, methyl and other functional groups), have been synthesized and are available commercially. Cyclodextrins present a conical shape with a hydrophobic inner cavity and the hydroxyl groups of the glucoses placed on both external sides, which confer them a polar character, and therefore aqueous solubility, while their interior is mostly hydrophobic. The twice-methylated derivative of β-cyclodextrin, namely, heptakis(2,6-di-*O*-methyl)-β-cyclodextrin (DIMEB), has been shown to induce a full rupture of Pluronic micelles [4-6], and the break-up mechanism is very selective on the type of cyclodextrins [5]. In addition, cyclodextrins are well known to form complexes with a wide range of drugs [7], therefore the ternary systems of

DIMEB, Pluronic and drug leads to interesting 3-way competitive interactions [6]. Finally, structures and interactions can be modulated by temperature and pH, as we show in this paper. Responsiveness to these various triggers and parameters, and thus the ability to tune the self-assembly structures, make these systems good candidates for the development of smart drug carriers for controlled release.

In this work, we study the drug-loading effect of two selected drugs of different structures on Pluronic F127 micelles: pentobarbital (PB) and naproxen (NP), both as sodium salts. We then examine the effect of the loaded drugs on the ability of DIMEB to break-up F127 micelles at various pH (pH 1, pH 13 and natural pH) and temperatures (25°C and 37°C) and provide some clues on the mechanism of break-up.

2. Experimental

2.1. Materials. Pluronic copolymer F127, comprising a central block of 65 PPO units and two side-blocks of PEO (100 units each), was obtained from Sigma-Aldrich UK. Heptakis(2,6-di-*O*-methyl)-β-cyclodextrin, naproxen sodium salt and pentobarbital sodium salt were obtained from Sigma-Aldrich UK.

2.2. Methods. Small-angle neutron scattering (SANS) measurements were carried out on LOQ at the ISIS facility (Rutherford Appleton Laboratory, Didcot, UK). The instrument uses incident wavelengths from 2.2 to 10.0 Å, sorted by time-of-flight, and a sample-detector distance of 4.1 m, providing access to scattering vectors q from 0.009 to 0.287 Å^{-1}.
Kinetic SANS measurements were carried out on the D22 instrument at the ILL using a stopped-flow equipment [8]. The wavelength was set at 6 Å, the sample-to-detector distance was 4 m, providing a q-range from 1.2×10^{-2} Å^{-1} to 0.26 Å^{-1}. The scattering measurements were made with an acquisition time for each frame n of t_n following a geometric series during 10 minutes:

$$t_n = a^{n-1}t_1,\ T_n = \frac{1-a^n}{1-a}t_1$$

where T_n is the accumulated time after mixing, with t_1 = 100 ms and a = 1.1. Sixty eight frames were measured for a total time of 651.7 s, after which 20 additional frames were measured for 1 s exposure each. Additional details can be found here [9].

3. Results and discussion

We first studied the effect of the pH (1, 13 and natural pH) and temperature (25°C and 37°C) on the properties of the micelles loaded with 1% PB or NP.

Figure 1A shows the scattering from solutions of 5 wt% F127 micelles at 25°C in the presence of 1 wt% PB at pH 1 and pH 13 (data at natural pH have been published elsewhere [6]), and also at 37°C and natural pH. The scattering from F127 shows typical features of spherical micelles [10]. At pH 13, the addition of 1 % PB hardly affects the scattering, as it had also been observed at natural pH (ca. 9) [6]; only a slight shift of the curve to higher q-values and decrease of the low-q intensity suggest a small decrease in size and increase in inter-micellar repulsions, possibly due to charge, since at both pH values the drug is dissociated. Overall however, although PB partitions within the micelles [6], it hardly affects their shape and size, thus suggesting a rather superficial localisation at the micelle/water interface. At pH 1 instead (the drug is in its molecular form), the scattering increases dramatically and stronger features are seen in the scattering pattern, showing that the partitioning of PB induces micellar growth. At a higher temperature of 37°C (and natural pH), the micelles are larger (Figure 1A), with a higher aggregation number [4], and, so are the micelles in the presence of PB; however, as seen at 25°C and both at natural pH and pH 13, the addition of PB has a very limited impact on micellar size.

In Figure 1B, the effect of adding DIMEB to the same solutions is shown. At pH 1 and 13 (at 25°C), the addition of 9% DIMEB leads to a complete rupture of the micelles, as had been observed at natural pH, both in the presence and absence of 1% PB [6]. At 37°C instead, the disruption of the micelles is quite limited; they decrease in size but are not ruptured (as inferred from the persisting strong features in the scattering pattern). Interestingly, while the micelles with 1% PB at pH 1 and 25°C are larger than at 37°C, they are ruptured much more substantially by the same amount of DIMEB. In other words, the temperature more efficiently opposes demicellisation than low pH. However, we had also shown that with a higher amount of PB (2%), the effect of DIMEB was substantially weakened at 25°C and pH 9 [6]. There is a clearly a balance of repulsive-attractive forces involved, which either hold the micelles together or oppose micellisation, and are modulated by the partitioning of the drug, the interaction of both the polymer and the drug with DIMEB, pH and temperature. Both pH and temperature affect the affinity between the various players (polymer chains, drug and cyclodextrins).

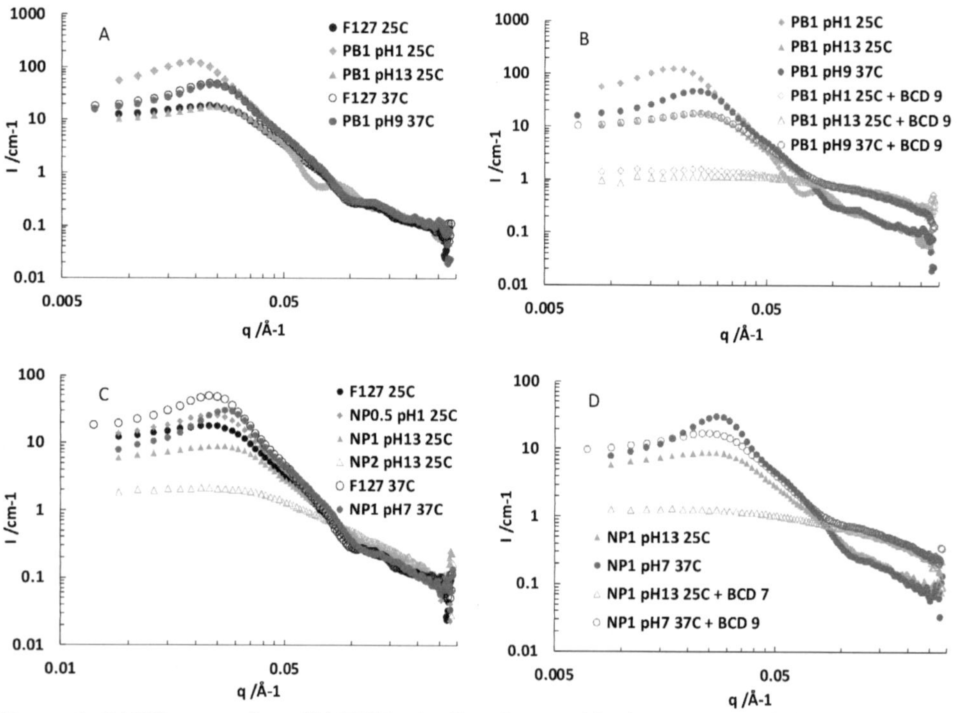

Figure 1. SANS curves from 5% F127 micelles alone and in the presence of either PB or NP, at various pH (pH 1, pH 13 and natural pH) and temperatures (25 and 37°C): **A.** Micelles with 1% PB at pH 1 and pH 13 (25°C), and at pH 9 (37°C). **B.** Micelles with 1% PB as in A and in the presence of 9% DIMEB. **C.** Micelles with 1% NP at pH 1 and pH 13, 2% NP at pH 13 (25°C), and with 1% NP at pH 7 (37°C). **D.** Micelles with 1% NP at pH 13 (25°C) and with 7% DIMEB; and at pH 7 (37°C) with 9% DIMEB.

We next studied the effect of NP on the micelles (Figure 1C and 1D). At pH 13 (25°C), the effect is similar than at neutral pH [6] (Figure 1C): adding NP (1 and 2%) decreases the size of the micelles, which is a clear sign of the partitioning of the drug inside the micelles, leading to charge repulsion and thus to a reduction of the aggregation number. Instead, at pH 1, adding 0.5% NP induces a growth of the micelles. At this pH the drug is in molecular form, therefore likely to partition deeper inside the micelles, and thus leading to a very different outcome than when in its charged form. At 37°C and neutral pH, the effect is similar to 25°C (and pH 7 or 13): micelles decrease in size when adding NP (but overall they are larger to start with at higher temperature).

As shown in Figure 1D, adding 7% DIMEB to the solutions containing 1% NP at pH 13 leads to full de-micellisation (as had also been observed at pH 7 [6]). Instead, it had been observed that with a higher amount of drug loaded (2% NP) and pH 7, the addition of the same amount of DIMEB induced an increase in micellar size [6], thus DIMEB offsets the disruptive effect of the partitioning of NP inside the micelles, possibly because of the preferential interaction of DIMEB with the drug, 'snatching' drug molecules from the micelles to form a β-CD drug complex [6]. However, at 37°C and pH 7 (Figure 1D), as also observed above with PB, the effect of DIMEB is substantially weakened by the presence of 1% NP, since the addition of 9% DIMEB leads only to a partial reduction in micellar size but no break-up.

Two possible mechanisms have been proposed to explain the modulation of the disruptive effect of DIMEB in the presence of drugs [6]:

- Hypothesis (1): competition between drug and F127 for complexation with CD reduces the availability of CD molecules to rupture the micelles
- Hypothesis (2): the partitioning of the drug affects the ability of the micelles to interact with F127.

More insight into the mechanism of rupture and the rules controlling it has been obtained for a range of compositions by measuring the binding constant of the drugs to the micelles and to DIMEB [6], and provides scope for a more in-depth discussion of the mechanisms.

The most widely invoked mechanism reported in cyclodextrins-polymer interaction is pseudopolyrotaxane (PR) formation, a type of inclusion complex involving the cyclodextrins threading onto the polymer backbone [11], acting as a rotor. Our first hypothesis was that this type of structure could be at the origin of micellar break-up [4]: by threading onto the Pluronic chains, and preferentially locating on the hydrophobic PPO block (the micellar core), the cyclodextrins would solubilise the micelles [4, 12]. However, our recent NMR studies [13] have revealed that the CH_3 group of PPO interacts weakly with the inner cavity protons H_3 and H_5 of the cyclodextrins, as well as with the CH_3 group in position 6. We proposed instead the formation of a "*loose*" inclusion complex, between DIMEB and the lateral CH_3 group of the PPO group from the polymer [13]. This would also justify the highly discriminative nature of the rupture mechanism on CD substituents [5,13].

More compelling evidence was recently provided by time-resolved SANS (TR-SANS) measurements on the naked micelles [9]. We report here new TR-SANS results on the drug-loaded micelles.

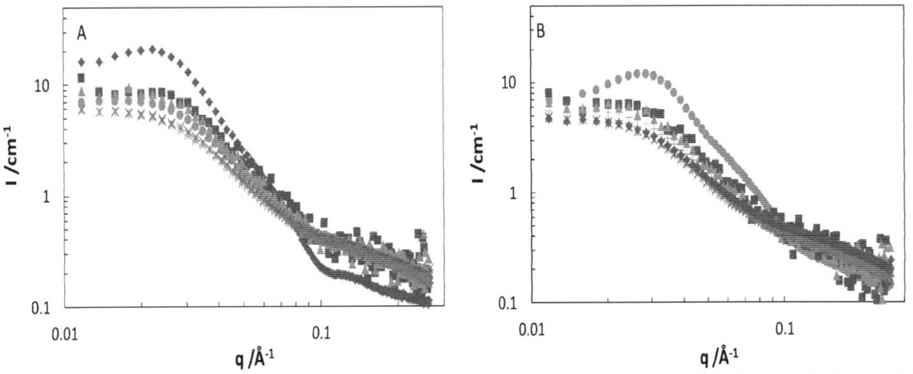

Figure 2. Small-angle neutron scattering data from 4 wt% F127 micelles with drug (◆), either 1% PB (A) and 1% NP (B) in the presence of 5% DIMEB at different time points after mixing: after 0.1 s (frame 1, ■), frame 1.5 s (frame 10, ▲), 111.1 s (●), 622.1 s (frame 68, ×), 642.1 s (frame 88, ✳).

Figure 2A and B show the effect of adding 5% DIMEB to 4% F127 solutions containing either 1% PB or NP. The first frame, taken less than 100 ms after mixing, reveals a drastic, instantaneous change in the scattering pattern, which reflects a shrinking of the micelles, as described above. This clearly demonstrates that the disruption of the micelles by DIMEB occurs in less than 100 ms. The threading of cyclodextrins onto linear hydrophobic polymers to form PRs is a lengthy process (over 10 s [14]); therefore, these measurements unequivocally rule out the formation of a PR as the mechanism of micellar break-up, as previously demonstrated for the free micelles [9,13]. Interestingly however, in the presence of NP and PB, scattering patterns measured at increasing times show a slight evolution of the disruption of the micelles. This had not been observed in the absence of drugs [9] and thus hints to a slightly different (marginally slower) mechanism of rupture in the presence of the drug, possibly involving loaded drug-cyclodextrin interaction as an intermediate step in the process, occurring over a few minutes. In the absence of drug, no changes were detected after 100 ms [9]. After ca. 620 s (ca. 10 min), the pattern has stabilised and overlaps perfectly with the equilibrium mixture of Pluronic, drug and DIMEB (Figure 2B).

4. Conclusion

We have shown that DIMEB can be used as a trigger to rupture Pluronic micelles. The partitioning of drugs modulates this behaviour, as well as the temperature and the pH. While the detailed mechanism is still under study, we have unambiguously ruled out the widely-invoked polypseudorotaxane formation as the mechanism of rupture and propose instead the formation of a loose complex between the polymer chains and the cyclodextrins. The same instantaneous micellar disruption reported before [9] is demonstrated here for the drug-loaded micelles, thus ruling out PR formation, however a further (slight) shrinking of the aggregates is observed here for the first time, suggesting a slightly different mechanism in the presence of drugs.

More insight into the balance of interactions controlling the break-up needs to be obtained in order to further develop this system as a smart drug carrier.

Acknowledgements. ILL and ISIS are acknowledged for the provision of neutron beam time. Richard Heenan and Isabelle Grillo are thanked for their help with the measurements at ISIS and ILL, respectively.

References

[1] Alexandridis P, Hatton T A 1995 *Colloids Surf.* A **96** 1-46
[2] Batrakova E V and Kabanov A V 2008 *J. Control. Release* **130** 98-106
[3] Szejtli J 1998 *Chem. Rev.* **98** 1743-54
[4] Joseph J, Dreiss C A, Cosgrove T and Pedersen J S 2007 *Langmuir* **23** 460-66
[5] Dreiss C A, Nwabunwanne E, Liu R and Brooks N J 2009 *Soft Matter* **5** 1888-96
[6] Valero M, Dreiss C A 2010 *Langmuir* **26** 10561-71
[7] Brewster M and Loftsson T 2007 *Adv. Drug Delivery Rev.* **59** 645-66
[8] Grillo I 2009 *COCIS* **14** 402-08
[9] Valero M, Grillo I and Dreiss C A 2012 *J. Phys. Chem.* B **116** 1273-81
[10] Pedersen J S and Gerstenberg MC 1996 *Macromol.* **29** 1363-65
[11] Dreiss C A, Newby F N, Sabadini E and Cosgrove T 2004 *Langmuir* **20** 9124-29
[12] Harada A, Li J and Kamachi M 1992 *Nature* **356** 325-27

Journal of Physics: Conference Series **549** (2014) 012011 doi:10.1088/1742-6596/549/1/012011

Confined dynamics in poly(ethylene terephthalate): a coherent and incoherent neutron scattering study

Alejandro Sanz[1], Aurora Nogales[1], Inés Puente-Orench[2,3], Mari-Cruz García-Gutiérrez[1], Javier Campo[2], Wolfgang Haussler[4], Michelina Soccio[5], Nadia Lotti[5], Andrea Munari[5] and Tiberio A. Ezquerra[1]

[1]Instituto de Estructura de la Materia, IEM-CSIC, Serrano 121, 28006 Madrid, Spain
[2]Institut Laue-Langevin, BP 156-38042 Grenoble Cedex 9, France
[3]Instituto de Ciencia de Materiales de Aragón, ICMA-CSIC, Pedro Cerbuna 12, 50009 Zaragoza, Spain
[4]Forschungs-Neutronenquelle Heinz Maier-Leibnitz (FRM II), Garching, Germany
[5]Dipartimento di Ingegneria Civile, Ambientale e dei Materiali-Universit a di Bologna, Via Terracini 28, 40131 Bologna, Italy

E-mail: alejandro.sanz@csic.es

Abstract. We show that the combination of dielectric relaxation with neutron spin echo and incoherent neutron backscattering measurements performed in deuterated and protonated poly(ethylene terephthalate) suggests that the intrinsic dynamics of semicrystalline polymers occurs in an homogeneous scenario, similar to that valid to describe the dynamics of totally amorphous polymers. The quasielastic neutron scattering data are satisfactorily described by a theoretical model that considers that the proton mobility follows a random jump-diffusion in a restricted environment.

1. Introduction

The physics of soft matter in confining media has attracted plenty of attention during the last decades for its implications on nanoscience, nanotechnology and fundamental knowledge of condensed matters [1]. The α relaxation mode is directly linked to the glass formation and becomes critically slow as temperature approaches the glass transition [2,3]. Polymers show the ability of self-assembling mainly through ordering processes like crystallization of homopolymers or phase separation of di-block copolymers. In both cases, hierarchical structures are generated at different length scales and normally one of the phases is confined by the surrounding media. The α relaxation in semicrystalline polymers has been widely investigated as a paradigmatic case study of how the geometrical confinement at the nanoscale influences the primary relaxation [4]. Poly(ethylene terephthalate) (PET) is a commodity polymer that, depending of several experimental factors, may present a variable degree of crystallinity, from the fully amorphous state to a 40 % approximately. Here, we have investigated the amorphous chains dynamics in a semicrystalline sample of PET by quasielastic neutron scattering (QENS), elucidating in this way how the confining crystals that surround the amorphous domains affect their mobility. To tackle this problem, quasielastic coherent Neutron Spin Echo (NSE) and incoherent neutron backscattering (BS) have been used on deuterated and protonated PET respectively.

2. Experimental section

Deuterated and semicrystalline poly(ethylene terephthalate) d-PET (degree of crystallinity X = 32 %) was specifically utilized for the NSE measurements. For the BS measurements, films of protonated poly(ethylene terephthalate) (X = 33 %) were used. Differential scanning calorimetry was utilized to estimate the glass transition temperature Tg of semicrystalline PET, located at 348 K.

2.1. Coherent scattering: Neutron spin echo

By using neutron spin echo spectroscopy (NSE), we report on a direct observation of the intermediate scattering function, or the dynamic structure factor, on a fully deuterated sample of poly(ethylene terephthalate) [5]. The dynamics of the density fluctuations in the proximity of the structure factor maximum is governed by the α-relaxation. Considering that the functional form for this process is a KWW (Kohkrausch-Williams-Watts) function [6], then for the NSE intermediate scattering function the following time dependence can be written:

$$\frac{S_{pair}(Q,t)}{S_{pair}(Q,0)} = f_Q(T)\exp\left[-\left(\frac{t}{\tau_s(T)}\right)^{\beta}\right], \tag{1}$$

where $f_Q(T)$ is related to the Debye-Waller factor, $\tau_s(T)$ is the structural or α relaxation time and β is the stretching parameter [14]. Neutron spin echo (NSE) experiments were performed at the RESEDA instrument at the FRM II Munich, Germany, at a Q value for the scattering vector of 1.38 $Å^{-1}$.

2.2. Incoherent scattering: Neutron backscattering

The scattering law for quasielastic processes is generally expressed by a lorentzian function with a half width at half maximum that provides the time scale of the microscopic dynamics. The lorentzian function is modulated by the term (1-EISF), where EISF is the elastic incoherent structure factor, in such a way that the model used to fit the experimental data is the following:

$$S_{inc}(Q,\omega) = EISF(Q)\delta(\omega) + \left[(1-EISF(Q))\frac{1}{\pi}\frac{\Gamma(\omega)}{\Gamma^2(\omega)+\omega^2}\right]. \tag{2}$$

The convolution of this model function to the experimental resolution data for each value of Q is what we actually fit to the experimental data in order to extract the half width at half maximum Γ (time scale of the motion) of the quasielastic component and the elastic incoherent structure factor EISF. BS experiments were performed at the backscattering spectrometer IN10, at the ILL (Grenoble, France). The incident neutron wavelength was set at 6.28 Å and the Q range was extended from 0.5 to 2 $Å^{-1}$, with an energy resolution at half maximum of 0.45 μeV.

3. Results and discussion

The dynamic behavior of semicrystalline PET has been investigated above its T_g by carrying out in parallel NSE and BS measurements. To illustrate the response of semicrystalline PET in time and frequency domain, figures 1 and 2 shows NSE and BS data respectively at 500 K.

Figures 1 and 2 nicely describe how the decay of the density correlations in the time domain, characterized by an exponential decay function, is on the contrary represented by a lorentzian function in the frequency domain [6]. The NSE data have been described by using the equation (1) where $f(Q)$, τ and β where adjustable parameters. The values of τ are included in the relaxation map represented in figure 4, supporting our initial assumption that the dynamics observed in NSE is truly the α relaxation [5]. While the stretching parameter β for the α relaxation in semicrystalline PET, for instance obtained by dielectric spectroscopy, show values around 0.2, here NSE data present decay functions with values of β equal to 0.5. Values of the parameter β close to 0.5 are typical for fully amorphous polymers. Consequently, we can propose that the dynamic scenario for the α-relaxation, studied by NSE, of a semicrystalline polymer is that of the homogeneous case. In order to extract additional information,

BS data were analyzed by using the formalisms described in equation 2, where the spectrum contains an elastic peak and the corresponding quasielastic broadening. As an example, figure 2 shows the best fit of the experimental BS data to equation 2 at 500 K and 1.45 Å^{-1}. Following this approach, a good description of the data was obtained also at 400 and 450 K in a Q range between 0.5 and 2 Å^{-1}. Unique information on the geometry of the motion is obtained through the Q evolution of the so called elastic incoherent structure factor EISF (ratio of the elastic intensity to the total scattered intensity). Two different approaches have been followed to interpret our EISF data, one in which the mobile species diffuse in a restricted environment defined by impermeable walls (DRE model) [7], and another where the confining elements behave like soft and permeable walls (Gaussian Model, GM) [8]. Fits of the GM and DRE models to our EISF experimental data are included in figure 3 as dashed and continuous lines respectively.

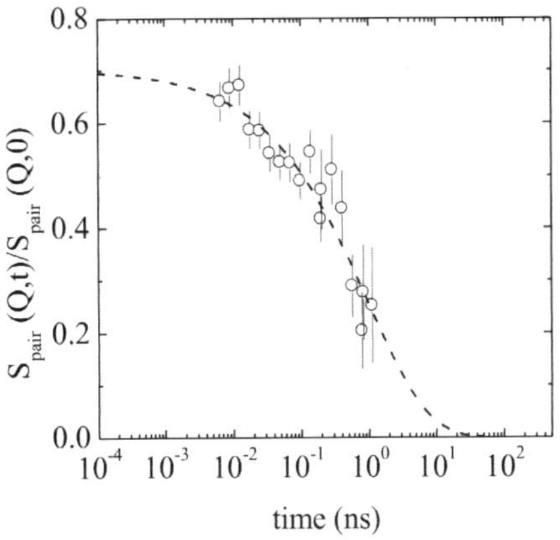

Figure 1. Dynamics of semicrystalline PET at 500 K. NSE relaxation decay at 1.38 Å^{-1}. Dashed line corresponds to the best fit of the experimental data to a KWW function.

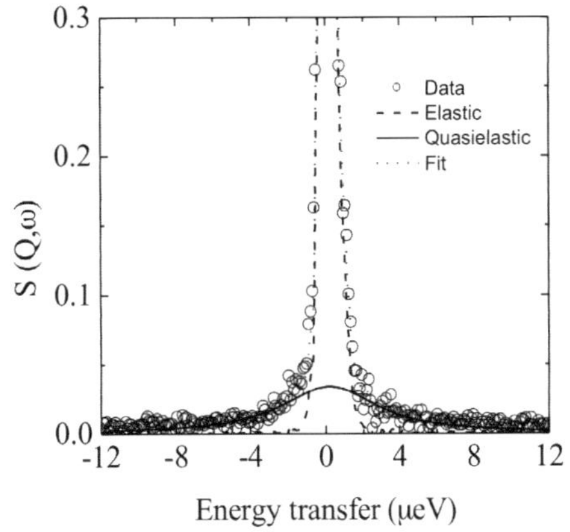

Figure 2. Dynamics of semicrystalline PET at 500 K. BS spectrum at 1.45 Å^{-1}.

If one checks carefully the fits in figure 3, it seems that a better description of the experimental data is received when the DRE model is considered. Under the framework of this model, the confining size remains unchanged for the whole temperature at a constant value of 6 Å, while the fraction of mobile protons increases with temperature. From the half width at half maximum Γ of the quasielastic component is possible to extract the time scale of the underlying motions. A detailed description of the data analysis procedure to estimate the relaxation time of particles diffusing in a restricted environment can be found elsewhere [9]. The relaxation times derived from both NSE and BS data are collected in figure 4, together to dielectric spectroscopy (DS) data. In figure 4, we have only included the 500 K data from the BS measurements considering that at lower temperatures the α relaxation time scale should fall far away from the IN10 time window. The dotted line in Fig. 4 corresponds to the empirical Vogel-Fulcher-Tamman VFT equation that nicely reproduces the cooperative nature of the α relaxation of glass forming systems. We see that the NSE data (⊖) fall into the VFT line, indicating that the collective dynamics observed through the coherent dynamic structure factor is actually the α relaxation. Interestingly, the data at 500 K obtained from the backscattering data almost coincide with NSE values.

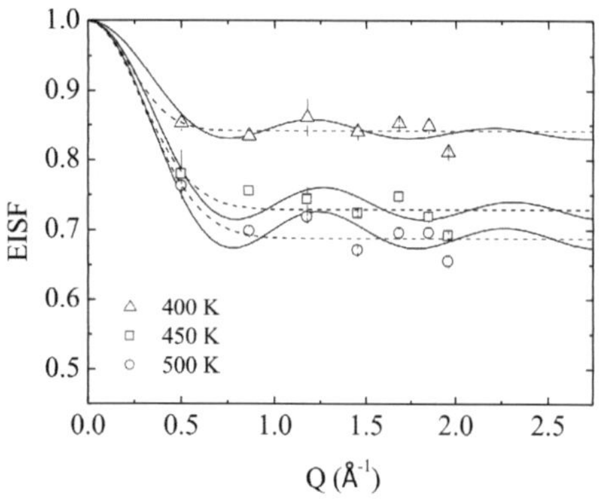

Figure 3. EISF data for semicrystalline PET. Solid lines represent the DRE model. Dotted lines correspond to the GM model. Read text for details.

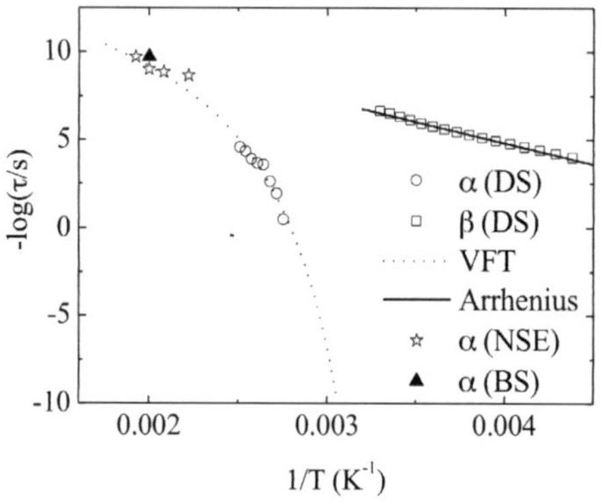

Figure 4. Relaxation times as a function of reciprocal temperature for semicrystalline PET. The graph collects data from different techniques.

Conclusions

The dynamics of semicrystalline PET at the mesoscopic level has been studied by quasielastic neutron scattering (QENS) using Neutron Spin Echo (NSE) and Neutron Backscattering (BS) techniques. Our EISF data have been modelled assuming that the scattering particles diffuse within a restricted environment (confining distance 6 Å). In addition to that, we observe an increase of the fraction of the mobile protons when temperature rises up. By doing NSE measurements above the T_g of deuterated semicrystalline poly(ethylene terephthalate), we were able to have access to the dynamics governed by the α relaxation, revealing that the intermolecular cooperativity is expected to be rather similar in both amorphous and semicrystalline polymers, suggesting that the dynamics of semicrystalline polymers occurs in an homogeneous scenario, similar to that valid to describe the dynamics of totally amorphous polymers [5].

To conclude, the inclusion of the BS data at 500 K into the relaxation map of semicrystalline PET indicates that the dynamics observed in this spectrometer is connected to the genuine α relaxation of semicrystalline PET. We demonstrate that the combination of coherent (NSE) with incoherent (BS) neutron scattering allows a complete description of the confined dynamics on a paradigmatic semicrystalline polymer like poly(ethylene terephthalate).

Acknowledgments

This work has been funded by the Spanish MINECO (MAT2009-07789, MAT2011-23455 and MAT2008-03232). A. S. also acknowledges the Spanish MINECO for financial support. The authors thank the Institut Laue-Langevin for beam time, M. A. González for technical assistance during experiments and Irene Calvo for helpful discussions.

References

[1] Frick B, Koza M and Zorn R 2003 European Physical Journal E **12** 3
[2] Angell C A 1995 Science **267** 1924
[3] Schönhals A and Kremer F 2002 Broad Band Dielectric Spectroscopy (Berlin: Springer)
[4] Ezquerra T A, Balta-Calleja F J, and Zachmann H G 1994 Polymer **35** 2600
[5] Sanz A et al. 2011 Macromolecules **44** 8124
[6] Bée M 1988 Quasielastic neutron scattering: principles and applications in solid state chemistry, biology, and materials science (Bristol and Philadelphia: Adam Hilger)

[7] Hall PL and Ross D K 1981 Molecular Physics **42** 673
[8] Volino F, Perrin J C and Lyonnard S 2006 Journal of Physical Chemistry B **110** 11217
[9] Nogales A et al. 1999 Macromolecules **32** 2301.

Journal of Physics: Conference Series **549** (2014) 012012 doi:10.1088/1742-6596/549/1/012012

Polymer Diffusion in Microgels with Upper Critical Solution Temperature as Studied by Incoherent Neutron Scattering

D Serrano Ruiz, P Alonso Cristobal, M Laurenti, J Rubio Retama, E Lopez-Cabarcos

Department of Physical-Chemistry II, Faculty of Pharmacy, University Complutense of Madrid, 28040 Madrid, Spain. Fax: 34-91394 2032; Phone: 34-91394 1751

E-mail: cabarcos@farm.ucm.es

Abstract. Poly(acrylic-acrylamide) interpenetrated microgels present continuous phase transition from collapsed to swollen state around 42 ºC. The upper critical solution temperature (UCST) of this polymeric system has prompted scientists to consider them candidates for its use in biological applications such as smart drug delivery devices since the swelling of the polymer matrix would permit the release of the drug previously entrapped within the microgels. In these systems the increment of the temperature can break inter-chain interactions, mainly hydrogen bonds, which reduce the elastic tension that stabilizes the microgel, favoring the polymer swelling. The microgel molecular dynamics at the UCST can be investigated using Incoherent Elastic (IENS) and Quasielastic Neutron Scattering (IQNS). From the analysis of the IQNS data we obtained that the diffusion coefficient of the polymer segments depends on the composition of the interpenetrated matrix. Thus, at room temperature, microgels with a polymer composition of 50% of each component present a diffusion coefficient $1 \cdot 10^{-12}$ m^2/s, while for the microgels formed by only one component the diffusion coefficient is 5.10^{-10} m^2/s. This huge difference in the diffusion coefficient is conspicuously reduced when temperature increases, and we attribute this effect to the breaking of the inter-chain interaction. By means of FTIR-ATR analysis we have identified the groups that are involved in this phenomenon and we associate the breaking of the polyacrylic-polyacrylamide interactions with the swelling of the microgels.

1. Introduction

Microgels based on PNIPAM have been extensively studied due to their ability to swell at low temperature and to collapse when temperature increases above 33 ºC[1-4]. It has been speculated using thermoresponsive microgels as drug delivery systems [5]. This would require an opposite behavior to that of PNIPAM, that is, the microgel has to remain collapsed at room temperature to trap the drug, and when increasing temperature, the microgel should swell and thus release the drug. This behaviour was found in poly(acrylic acid-acrylamide) interpenetrated microgels (PAAc/PAA) which present continuous phase transition from collapsed to swollen state around 42 ºC. The UCST of this polymeric system has prompted scientists to consider them candidates for its use as smart drug delivery devices since the swelling of the polymer matrix would permit the release of the drug entrapped within the microgels. However, unlike PNIPAM where the volume phase transition spans a range of only 2-3 degrees, the transition in PAA elapses at an interval of about 30 ° C.

2. Experimental

2.1. Materials and methods

The UCST microgels were prepared using the concentrated emulsion polymerization method. The concentrated emulsion was prepared with a low fraction of the oil phase, which constitutes the continuous phase [6]. We have also prepared microgels with 100% poly(acrylic acid) (PAAc) and 100% poly(acrylamide) (PAA). In order to obtain PAA microgels, 5 mL of an aqueous solution consisting of acrylamide (17mmol), 12.5 mg of bisacrylamide (BIS), and ammonium persulfate (25 mg) was added with a syringe to the continuous oil phase (750 μL of dodecane and 250 μL of Span 80). To obtain, PAAc microgels we added sodium acrylate (17 mmol) keeping the same amount of BIS and ammonium persulfate used for PAA microgels. The emulsion was homogenized by magnetic stirring and purged with nitrogen to remove residual oxygen. The polymerization was started by adding TEMED (63 μL) and after 1 h of reaction, the polymer was precipitated and washed with acetone several times. Microparticles were isolated by centrifugation (10,000 rpm) for 10 min at 10 °C. In order to produce interpenetrated microgels (IPNs), part of the monomers were substituted by a defined amount of linear poly(sodium acrylate) or polyacrylamide (giving as mmol of repeated units). The resulting solution together with the initiator and cross-linking agent were used as internal aqueous phase of the concentrated emulsion and polymerized as described above. The Table 1 shows the microgels composition.

Table 1. Concentration of each component in the dispersed aqueous phase (5 ml water) used in the production of the microgels. The continuous oil phase consists of 750 μL of dodecane and 250 μL of Span 80.

AA(mmol)	AAc (mmol)	Linear PAA (mmol)	Linear PAAc (mmol)	Sample name
17	0	0	0	PAA
12	0	0	5	PAA/PAAc (70/30)
10	0	0	7	PAA/PAAc (60/40)
8.5	0	0	8.5	PAA/PAAc (50/50)
0	10	7	0	PAA/PAAc (40/60)
17	0	0	0	PAAc

Methods. The microgel particles were studied using transmission electron microscopy (SEM) at an acceleration voltage of 20 kV in a JEOL JSM-6400. The ATR-FTIR spectra were taken with a Nicolet IR200 FTIR spectrometer equipped with a Peltier (Linkam Scientific). Background spectra were taken before sample spectra and subsequently subtracted. Specifically, for each sample containing 20% The neutron scattering experiments were performed at the Institut Laue Langevin in Grenoble using the high energy resolution (1 μeV) neutron backscattering spectrometer IN10.

2.2. The incoherent scattering function

The IQNS data were analyzed considering the molecular motion consist in a combination of a faster vibrational component with a mean square displacement $\langle u^2\rangle^{1/2}$ and a translational diffusion motion with diffusion constant D [7-8]. For Fickian diffusion the dynamic structure factor is a Lorentzian function, with a half-width at half-maximum $D.Q^2$. Thus, the scattering function can be written as,

$$S(Q,\omega) = \exp(-\frac{1}{3}\langle u^2\rangle Q^2)\delta(\omega) \otimes \frac{1}{\pi}\frac{DQ^2}{(DQ^2)^2+\omega^2} \quad (1)$$

where $\delta(\omega)$ is a delta function, and $\otimes$ is the convolution product in ω. The mean square displacement $\langle u^2\rangle^{1/2}$ can be derived from the Q^2 dependence of $S(Q,0)$ at fixed temperature, and D is obtained from the fitting of the quasielastic component with a Lorentzian function [7-8].

3. Results

Figure 1 shows a SEM image of the PAA/PAAc (70/30) microgels. As is shown in Figure 1 the polymerization of the concentrated emulsion produces microgel particles with diameters between 0.3 and 3 μm, which is approximately the size of the precursor emulsion polyhedral cells. The particles are spherical and the dispersity is low. With the aim of evaluating the UCST behavior of the microgels with different polymer composition, we study the increment of their volume as a function of the temperature and the results are depicted in Figure 2. This measurement was performed in water at pH 3 since at this pH the UCST present its maximum.

Figure 1. SEM micrograph of microgel particles.

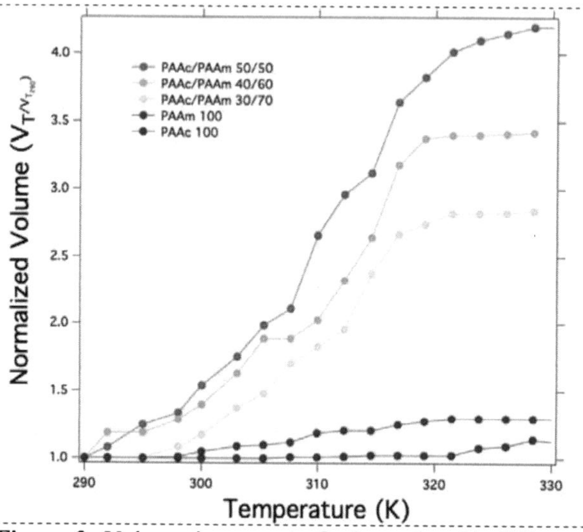

Figure 2. Volume increment of microgels with different polymer composition as a function of the temperature.

Diffusion motions of polymer chain segments can be investigated using inelastic neutron scattering since they are prominent in the quasielastic region. Furthermore, by dispersing the microgels in D_2O is possible to obtain enough contrast between the polymer network and the solvent. The elastic neutron-scattering function $S(Q, \omega=0)$ normalized to its value at 190 K, $S(Q, 0)_{190K}$ was measured as a function of the temperature with IN10 using the so-called "fixed elastic window" method. The measured incoherent elastic neutron scattering function as a function of temperature is shown in Figure 3. From this function, the apparent mean square displacement $<u^2>^{1/2}$ of the polymer chains can be obtained analyzing the Q dependence of $S(Q,0)$ at fixed temperature

$$\ln [S(Q,0)_T / S(Q,0)_{190K}] = -1/3 [\langle u^2(T)\rangle - \langle u^2(190 \text{ K})\rangle]\cdot Q^2 \quad (2)$$

From the analysis of the IQNS data (see Figure 4) we obtained that the diffusion coefficient of the polymer segments depends on the composition of the interpenetrated matrix. Thus, at room temperature, PAA/PAAc (50/50) microgels present a diffusion coefficient $1\cdot10^{-12}$ m^2/s, while for microgels formed by only one component the diffusion coefficient is 5.10^{-10} m^2/s in both cases, PAA and PAAc.

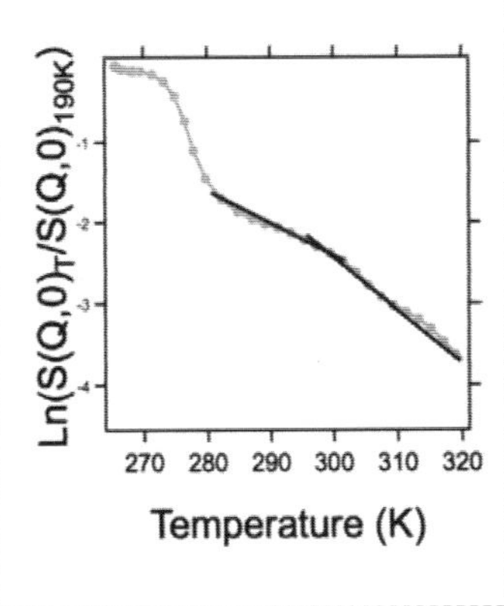

Figure 3. Normalized elastic scattering function during the heating of PAA/PAAc (50/50) microgels taken at $Q = 0.5$ Å^{-1}.

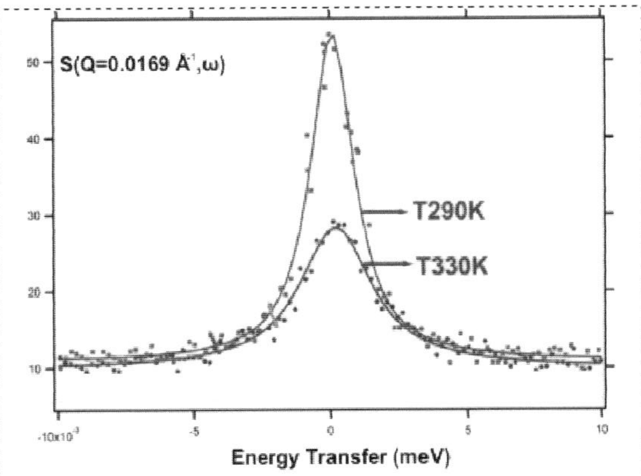

Figure 4. The $S(Q,\omega)$ function shows the different dynamics of the microgel networks in the swollen (330 K) and collapsed (290 K) states.

We have used IR spectroscopy to determine the interactions of the polymer matrix with the local environment. Specifically, we look for the change in the vibration bands of the amide and acrylic acid groups during the swelling process of the microgels. First, we collected the IR-ATR spectra of microgels dispersion of neat PAAc and neat PAA at 290 and 330 K. The PAA spectra present the following bands: 1607 cm^{-1} corresponding to the bending vibration amide II (C-NH2), 1635 cm^{-1} associated with the stretching vibration of the C=O···H-O-H , and 1670 cm^{-1} corresponding to the stretching vibration of the C=O. The PAAc presents a band at 1714 cm^{-1} corresponding to the stretching of the carbonyl group of the carboxyl acid. Thus, the peaks of the spectrum of the PAA/PAAc samples can be identified (see Figure 5). At 290 K, due to the formation of H-bonds, the band at 1730 cm^{-1}, corresponding to the stretching of the carbonyl group, shifts to 1714 and the band at 1635 shifts to 1610 cm^{-1}. When the temperature increases at 330 K, the intensity of the bands at 1607 and 1714 cm^{-1} decrease due to the H-bonds rupture. Thus, the swelling of the microgels seems to be governed by the breaking of hydrogen bonds above 300 K.

4. Conclusions

We have synthesized IPN microgels with different compositions using the concentrated emulsion polymerization method. These microgels present the greatest UCST when the matrix composition was PAA/PAAc (50/50). IQNS experiments permit determining the mean square displacement of polymer segments and their diffusion coefficient for the different microgel compositions. The FTIR spectrum seems to indicate that the swelling of the microgels is governed by hydrogen-bonds, which are broken above 300 K resulting in a reduction of the elastic tension of the polymer matrix

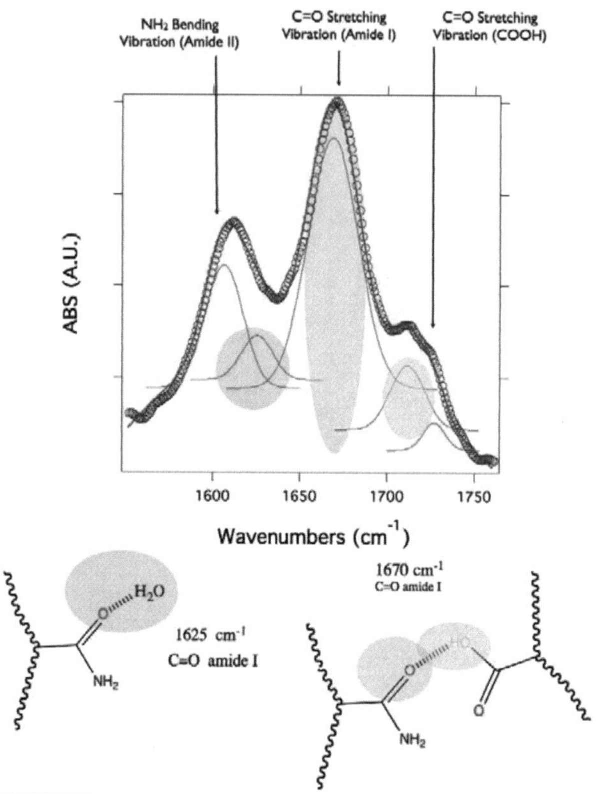

Figure 5. FTIR absorption spectra of the amide I, amide II and C=O stretching bands for PAA/PAAc microgels in the collapsed state (temperature = 25°C)

Acknowledgements
The authors acknowledge financial support from the Spanish Science and Innovation Ministry (grant MAT2010-15349), and from the COST Action CM1101.

References

[1] Fernández-Barbero A, Fernandez-Nieves A, Grillo I and Lopez-Cabarcos E 2002 *Physical Review E* **66** 051803
[2] Rubio-Retama J, Frick B, Seydel T, Stamm M, Fernandez-Barbero A and Lopez-Cabarcos E 2008 *Macromolecules* **41** 4739
[3] Rubio-Retama J, Zafeiropoulos N E, Serafinelli C, Rojas-Reyna C, Voit B, Lopez-Cabarcos E and Stamm M 2007 *Langmuir* **23** 10280.
[4] Pelton R H and Chivante P 1986 *Colloid. Surf.* **20** 247
[5] McPhee W, Tam K C and Pelton R H 1993 *J.Colloid Interface Sci.* **156** 24
[6] Echeverria C, López D and Mijangos C 2009 *Macromolecules* **42** 9118
[7] Bee M 1988 *Quasielastic Neutron Scattering*. Ed. Adam Hilger: Bristol
[8] Nogales A, Ezquerra T A, Batallan F, Frick B, Lopez-Cabarcos E and Balta-Calleja F J 1999 *Macromolecules* **32** 2301

Journal of Physics: Conference Series **549** (2014) 012013 doi:10.1088/1742-6596/549/1/012013

Microscopic dynamics of glycerol: a QENS study

A. Vispa[1], M. Rovira-Esteva[1], M.D. Ruiz-Martín[1], S. Busch[2], T. Unruh[3], L. C. Pardo[1], J. Ll. Tamarit[1]

[1]Grup de Caracterització de Materials, Departament de Física i Enginyieria Nuclear, ETSEIB, Universitat Politècnica de Catalunya, Diagonal 647, 08028 Barcelona, Catalonia, Spain

[2]Department of Biochemistry, University of Oxford, United Kingdom

[3]Lehrstuhl fur Kristallographie und Strukturphysik , Universität Erlangen, Germany

E-mail: alessandro.vispa@upc.edu

Abstract. We report on a quasielastic incoherent neutron scattering (QENS) experiment on liquid glycerol. QENS data were collected at the temperature T=380 K and with a resolution (FWHM) R=55μeV. The analysis of the quasielastic signal enables us to draw a consistent picture of the diffusive mechanism on a picosecond time scale and to compare with most recent models for glycerol dynamics. Model selection, performed with the fitting algorithm FABADA, gives us a preliminary description about the motions of the glycerol molecules in its liquid state.

1. Introduction

Glycerol $C_3H_5(OH)_3$ is one of the weak hydrogen-bonded systems composed of small and highly flexible molecules. The presence of three hydroxyl groups in the molecule gives rise to a particularly rich and complex conformational and structural behaviour in condensed phases. As a consequence glycerol is an excellent glass former and the nature of its glass transition has been the object of several experimental investigations [1-3, 8-10]. The understanding of the microscopic structure and dynamical processes in highly viscous liquids such as glycerol, aimed at a unified theoretical description of the system, has been the focus of a great deal of attention for many decades [1-9]. Parallel to the computation advances [10-12], the recent developments in reliable experimental techniques provide high-quality data. Despite such a high scientific attention, it is not yet clear which is the theoretical model that better describes the microscopic dynamics. Many experiments were done using neutron scattering [13-19], but obtaining the diffusion coefficients from the measured quasielastic spectra requires high experimental care. The technical difficulties intrinsic to this kind of experiments can be sensibly reduced exploiting the high signal-to-noise ratio attained by the time-of-flight spectrometer TOFTOF at the Forschungs-Neutronenquelle Heinz Maier-Leibnitz (FRM II, Garching bei München, Germany) [20].

Here, we report on a QENS experiment carried out on TOFTOF with the aim of measuring the incoherent quasielastic response of liquid glycerol. We also propose two alternative models to describe the microscopic dynamics of glycerol: diffusion+3D Brownian oscillator and a more restrictive model such as diffusion+rotation.

2. Data analysis and results from QENS spectra

Experiments were performed at TOFTOF (FRM II, Garching) using an incoming neutron beam with a wavelength of 6.0 Å. The Q-range investigated was $0.4<Q(\text{Å}^{-1})<1.4$. Data-reduction, including normalization and background subtraction was performed using the program FRIDA [21]. Spectra were obtained by regrouping the data into slices at constant Q [22]. The spectra were evaluated in the region of E = −0.5 meV to +0.5 meV where positive values for the energy transfer denote an energy gain of the neutron during the scattering process. The resolution was measured using a vanadium rod in the same instrumental configuration as the sample.

In order to analyze the data, the theoretical function must be convolved with the instrumental resolution $R(Q,\omega)$. When the width of the theoretical function Γ is close to the width of the energy bin ΔE, the theoretical function to be fitted can be reduced to a single point. In this case the broadening Γ obtained after the convolution process is greater than the real value [23]. This artefact is especially dangerous when fitting a diffusion process, since in this case Γ is proportional to the squared value of the momentum transfer Q. This causes that the value of Γ is bigger than the correct one in the low Q region. In other words, when using an energy bin ΔE too big related to the line broadening, a deviation from the Q^2 law is observed which is not related to any physical process, but to a numerical problem of the convolution. As stated in reference [24], in order to avoid this effect half the bin width must be smaller than the line width, i.e. $\Gamma \geq \Delta E/2$. In figure 1 we show the obtained values of Γ using a series of decreasing energy binnings (0.01, 0.005 and 0.001 meV). As it is clear in the figure using an energy bin too big could incorrectly lead to the conclusion that there is a level-off from the line broadening describing the diffusion at low momentum transfer. However, a fitting using smaller energy bins shows that, in fact, when the energy bin is small enough the Q^2 law is recovered, and the level-off at small Q is only a numerical artefact.

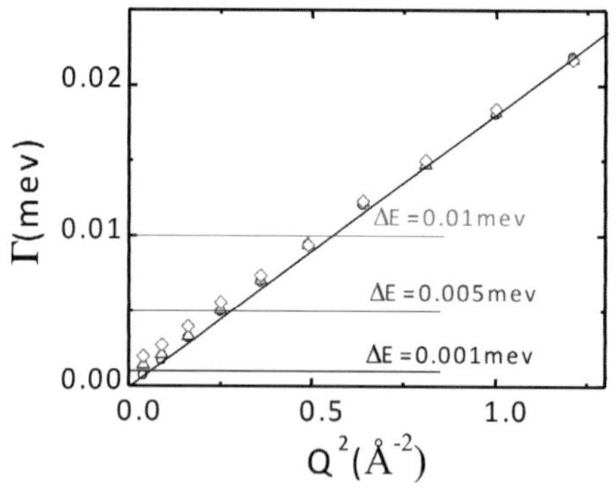

Figure 1.(colour online) Line broadening Γ as a function of the squared momentum transfer fitted using different energy binnings (diamonds for $\Delta\omega$=0.01 meV, triangles for $\Delta\omega$=0.005 meV and circles for $\Delta\omega$=0.001 meV). A series of lines showing the binning has been added for the sake of clarity.

In order to investigate which motions are taking place at a microscopic level, we have tested two models, both including a diffusion term and a confined motion at long times. First of all we checked a model where the particle executes a Brownian motion in an harmonic potential, the so called Brownian Oscillator (BO) [25]. The quasielastic scattering function is thus described in terms of a Lorentzian function related to the diffusion (D) with a broadening $\Gamma_D(Q)=DQ^2$, and the broadening related to the Brownian oscillator, Γ_{BO}. The scattering law can be written as:

$$S(Q,\omega) = L(\Gamma_D)\otimes\left(A_0(Q) + \sum_{m=1}^{\infty} A_m(Q)\frac{1}{\pi}\frac{m\cdot\Gamma_{BO}}{(m\cdot\Gamma_{BO})^2+\omega^2}\right) \tag{1}$$

$$\text{with} \quad A_0(Q) = e^{-Q^2\varepsilon^2} \quad \text{and} \quad A_m(Q) = (Q^{2m}\varepsilon^{2m})/m!$$

For the second model a rotational motion has been assigned to the confined motion. We will call this model Diffusion + Rotational motion (D+R) and the scattering law can be expressed as:

$$S(Q,\omega) = L(\Gamma_D) \otimes \left(A_0(Q) + \sum_{l=1}^{\infty} A_l(Q) \frac{1}{\pi} \frac{\Gamma_{rot}}{(\Gamma_{rot})^2 + \omega^2} \right)$$

$$\text{with} \quad A_0(Q) = j_0^2(QR) \quad \text{and} \quad A_l(Q) = (2l+1) j_l^2(QR) \tag{2}$$

where $\Gamma_{rot} = l(l+1)D_{Rot}$, and j_l are spherical Bessel function of l-th order.

Fittings were performed to the data at all Q-values simultaneously, in a way to fit, at once, both processes and avoid cumulative errors. In panels a and b in figure 2, we show two examples of the best fits to the data at two values of momentum transfer (Q=0.5 $Å^{-1}$ and at Q=1.1 $Å^{-1}$). As it can be seen both models are able to fit the data reasonably well at a first glance. The parameters coming from the best fit are, for the D+BO model: diffusion coefficient of $D=0.022 \cdot 10^{-4}$ cm^2s^{-1}, Γ_m=0.05 meV and ε=0.7Å and for the D+R model $D=0.025 \cdot 10^{-4}$ cm^2s^{-1}, rotational diffusion coefficient $D_{Rot}=0.054 \cdot 10^{-4}$ cm^2s^{-1} and the radius of gyration is R=1.2 Å. The values for the diffusion coefficients appear to be compatible with the values obtained from NMR experiments ($D=0.019 \cdot 10^{-4}$ cm^2s^{-1}), as stated in reference [26].

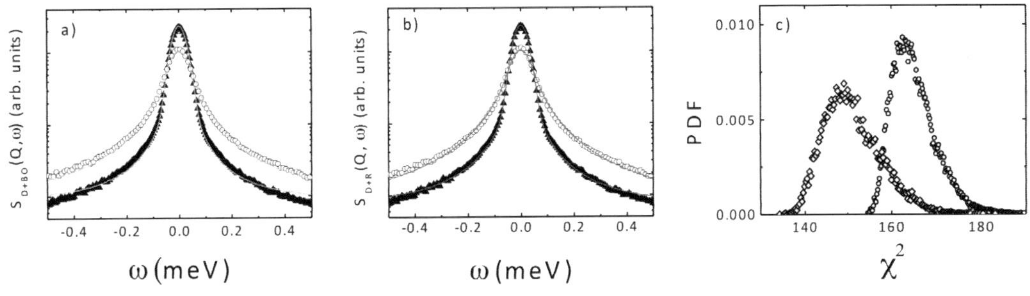

Figure 2. Experimental data together with the bests fit to two different models: Diffusion + Brownian oscillator (a) and Diffusion + Rotational motion (b) for Glycerol at T=380 K with a resolution of 55 μeV. For this two models we present two spectra and relative fits at the same Q values (triangles for Q=0.5 $Å^{-1}$ and circles for Q=1.1 $Å^{-1}$). (c) Probability density function for the figure of merit χ^2 for the two models (diamonds D+BO, circles for D+R model).

In order to determine in a quantitative way which model fits better the data we show the Probability Distribution Function (PDF) of the figure of merit χ^2, for the two proposed models: the Diffusion + Brownian oscillator motion is clearly preferred since any combination of parameters leads to a better fit.

3. Conclusion

We present in this work a method to quantify which model is the best describing the data, beyond the "good looking fit" criterion. The analysis of the data for glycerol leads us to the conclusion that a Brownian oscillator describes better the data of the confined motion than a simple rotation of the whole molecule.

Acknowledgments
This work was supported by the Spanish Ministry of Science and Innovation (Grant FIS2011-24439) and the Catalan Government (Grant 2009SGR-1251). We thank FRM-II for providing beam time on TOFTOF.

References

[1] Wuttke J, Hernandez J, Li G, Coddens G, Cummins H Z, Fujara F, Petry W and Sillescu H 1994 *Phys Rev Lett* **72** 3052; Wuttke J, Petry W, Coddens G, Fujara F 1995 *Phys Rev E* **52** 4; Wuttke J, Petry W and Pouget S 1996 *J Chem Phys* **105** 5177; Wuttke J, Chang I, Fujara F, Petry W *Physica B* 1997 **234-236** 431-432

[2] Towey J J, Soperb A K and Dougan L 2011 *Phys Chem Chem Phys* **13** 9397–9406

[3] Roland C M and Ngai K L 1997 *J Chem Phys* **106** 1187-1190

[4] Cunsolo A, Leu B M, Said A H, Cai Y Q 2011 *J Chem Phys* **134** 184502

[5] Scopigno T, Pontecorvo E, Di Leonardo R, Krisch M, Monaco G, Ruocco G, Ruzicka B and Sette F 2004 *Philosophical Magazine* **84** 1453–1461

[6] Padureanu I, Lechner R E, Aranghel D, Radulescu A, Desmedt A, Pieper J, Ion M 2002 *Appl Phys A* **74** [Suppl], 1342–1344

[7] Ediger M D, Angell C A, Nagel S R 1996 *J Phys Chem* **100** 13200-13212

[8] Debenedetti P G, Stillinger F H 2001 *NATURE* **410** 259-267

[9] Angell C A 2008 *MRS BULLETIN* **33** 544-555

[10] Chen B, Sigmund E E, Halperin W P 2009 *http://arxivorg/abs/cond-mat/0611529v1*

[11] Chelli R, Procacci R, Cardinia G, Della Valle G, Califano S 1999 *Phys Chem Chem Phys* **1** 871-877, Chelli R, Procacci R, Cardinia G, Califano S 1999 *Phys Chem Chem Phys* **1** 879-885

[12] Blieck J, Affouard F, Bordat P, Lerbret A, Descamps M 2005 *Chem Phys* **317** 253–257

[13] Franz H *et al* 1999 *Hyperfine Interactions* **123-124** 865–879

[14] Aranghel D *et al* 2010 *J OPTOELECTRON ADV M* **12** 4 953-957

[15] Kamiyama T, Miyamoto A, Kawamura J, Nakamura Y, Kiyanagi Y 1999 *Journal of Physics and Chemistry of Solids* **60** 1549–1552

[16] Larsson K F 1968 *Phys Rev* **167** 1 171-182

[17] Wuttke J, Chang I, Randl O G, Fujara F and Petry W 1996 *Phys Rev E* **54** 5364

[18] R Busselez *et al* 2011 *J. Phys.: Condens. Matter* **23** 505102

[19] Bee M 1988 *Quasielastic Neutron Scattering* Adam Hilger, Bristol and Philadelphia; Lovesey S 1986 *Theory of neutron scattering from condensed matter* Oxford Science Pubblication, (Clarendron press, Oxford)

[20] Unruh T, Neuhaus J, Petry W 2007 *Nucl Instrum Methods Phys Res A* **580** 1414–1422 and erratum **585** 201

[21] Guarini E 2003 *Scattering anelastico di neutroni e tecniche di analisi dati* ISI-INFM, XIV Scuola Nazionale di Fisica della Materia

[22] Wuttke J 2006 FRIDA (Fast Reliable Inelastic Data Analysis), http://apps.jcns.fz-juelich.de/doku/frida/start

[23] Busch S, Unruh T 2011 *J Phys: Condens Matter*

[24] Pardo L C, Rovira-Esteva M *et al* 2011 *Phys Rev E* **84** 046711; Pardo L C, Rovira-Esteva M, Busch S *et al* 2011 *J Phys: Conf Ser* **325** 012006

[25] Busch S 2012 *The Pico- to Nanosecond Dynamics of Phospholipid Molecules*, PhD Thesis, Technische Universität München

[26] Tomlinson D J 1972 *Molecular Physics* **25** 3

Short range order of CCl_4: RMC and MD Methods

A. Silva-Santisteban[1], A. Henao[1,2], S. Pothoczki[1,5], F. J. Bermejo[3], J. Ll. Tamarit[1], E. Guardia[2], G. J. Cuello[4] and L. C. Pardo[1]

[1]Grup de Caracterització de Materials, Departament de Física i Enginyieria Nuclear, ETSEIB, Universitat Politècnica de Catalunya, Diagonal 647, 08028 Barcelona, Catalonia, Spain

[2]Grup de simulació de material condensada per ordinador, Departament de Física i Enginyeria Nuclear, Universitat Politècnica de Catalunya, Barcelona, Spain

[3]CSIC- Department of Electricity and Electronics, UPV/EHU, Bilbao, Spain

[4]Institut Laue Langevin, 6 rue Jules Horowitz, 38042 Grenoble, France

[5]Institute for Solid State Physics and Optics, Wigner Research Centre for Physics, Hungarian Academy of Sciences, Konkoly-Thege Miklós út 29-33, 1121 Budapest, Hungary

E-mail: alvaro.silva-santisteban@estudiant.upc.edu

Abstract. The main objective of this paper is to present a method to fully determine the six degrees of freedom regarding position and orientation of a neighboring molecule around a central one, i.e. the $g(\mathbf{r}_{CM},\omega)$ distribution function. This is accomplished by completely determining the short range structure of liquid carbon tetrachloride, while employing results from two different methods, Molecular Dynamics (MD) [1] and Reverse Monte Carlo (RMC) [2]. Exclusively, the structural ordering of the first four molecules will be detailed.

1. Introduction

Due to the unavailability of a consensus on how to extract the detailed intermolecular structure of liquids made up of symmetric top molecules, we have sought to provide a method as to completely determine the structure of carbon tetrachloride (CCl_4), one of the first and most studied such molecular liquids [3-10].

Since this method is able to detect the slightest change in molecular ordering, we have also used it to compare the molecular orderings obtained both by MD and RMC [5, 7] methods. The former is a simulation requiring *a priori* information concerning intermolecular interaction potentials, whereas the latter method is a maximum entropy analysis of experimental diffraction patterns which needs no initial information aside from macroscopic density, as the most probable molecular ordering is obtained directly from experimental measurements.

2. Experiment

Neutron diffraction experiments were performed at the two-axis diffractometer D1B at Institut Laue Langevin (ILL) using a wavelength of 1.28 Å. Standard and Placzek corrections were performed to the obtained data. The chemical CCl_4 with purity higher than 99% was obtained from Across and was used as such.

3. Structural ordering

3.1. Positional ordering

Firstly, to describe the positional ordering a coordinate system has been defined on a central molecule. The origin is at the center of mass (CM), the z-axis has been placed along a carbon chloride (C-Cl) bond, the x-axis is co-planar with the z-axis and another bond, and finally the y-axis is the cross product of the previous two axes.

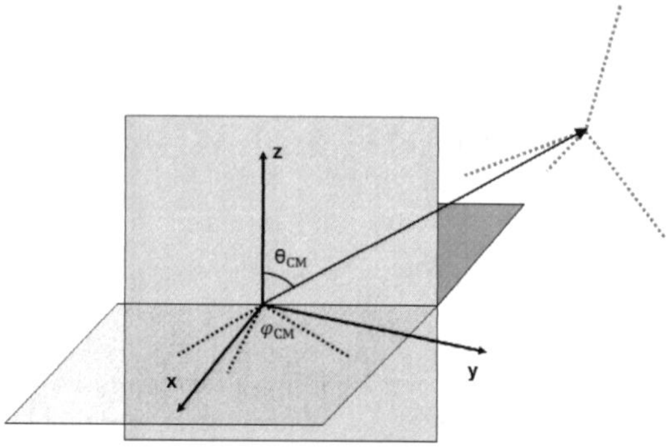

Figure 1. Using the newly defined coordinate system, a surrounding molecule's center of mass is determined by spherical coordinates (θ_{CM} and Φ_{CM}).

The 3D probability distribution graphs plotting the spherical coordinates (cos θ_{CM} - Φ_{CM}) of the positions of the first four adjacent molecules are displayed in figure 2, which show the areas of highest probability of a molecular position.

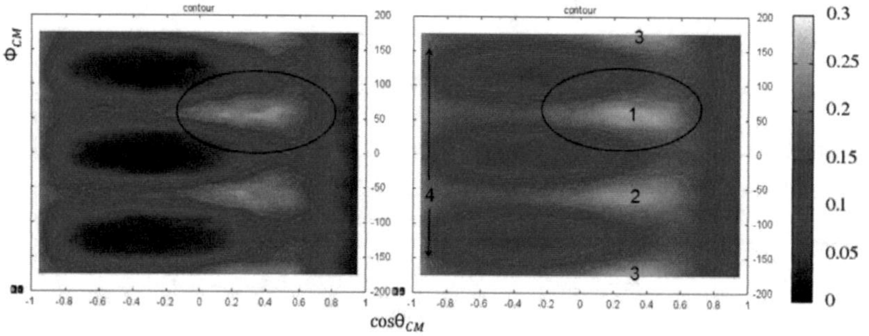

Figure 2. Graph from MD data is shown left and RMC on the right. Numbers are placed upon each probable position of a neighboring molecule.

The neighboring molecules present a tetrahedral arrangement around a carbon tetrachloride molecule. This central tetrahedral molecule has a neighboring molecule upon each of its 'faces', as can be seen in figure 3.

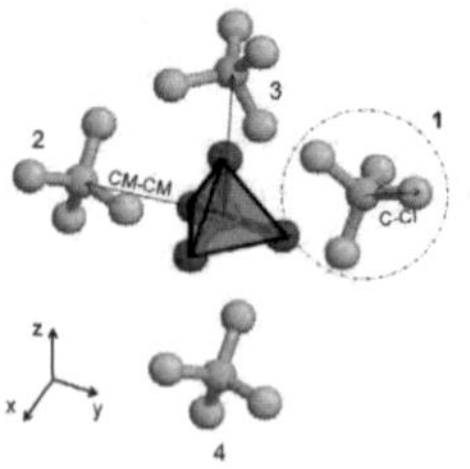

Figure 3. Most probable arrangement of four neighboring molecules around a central one.

3.2. *Orientational ordering*

In order to gather the orientational ordering of the first four molecules, the data of the molecule labeled as one (1) in figure 3 has been selected for further analysis and through symmetry the three remaining orientational configurations are deduced. In figure 4 we plot the probability map of the first two Euler angles in the ZY'Z'' convention (φ, θ, Ψ angles respectively) related to the orientation of any C-Cl vector of a neighbor molecule in regard to the central one.

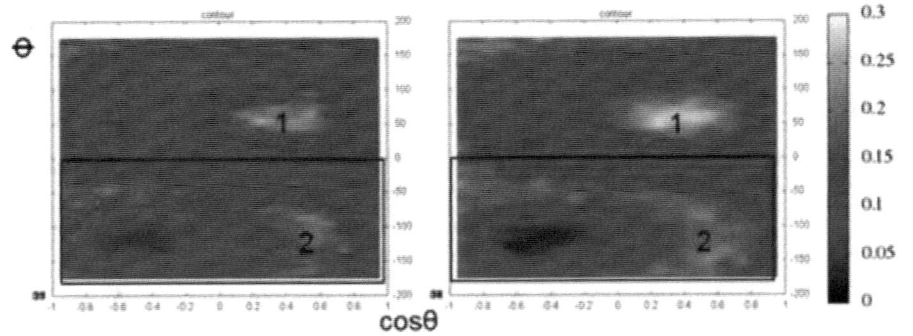

Figure 4. Two main high-probability orientational tendencies can be observed in the cosθ-φ graph. MD (left), RMC (right).

The area numbered as 1 in the figure at approximately cosθ=0.4, φ=60 is related to a C-Cl vector which is pointing opposite and perpendicular to a face of the central molecule. The other somewhat undefined area (2) which surrounds a point at about cosθ=-0.4, φ=-60, is caused by the three other C-Cl vectors, which are not perpendicular to the face of the tetrahedron. These results agree with previous studies of CCl_4, for which the face to face configuration is the most probable for closest molecules [4].
The remaining Euler angle ψ concerns the probability of a rotation around each C-Cl vector of the neighboring molecule: either around the C-Cl vector perpendicular to the face of the central tetrahedral molecule (area (1) in figure 4) or around the remaining three C-Cl vectors (area (2) in figure 4). Therefore, in order to further analyze these two different orientations, the cosθ-Ψ graph concerning each area of the graph above has been plotted.

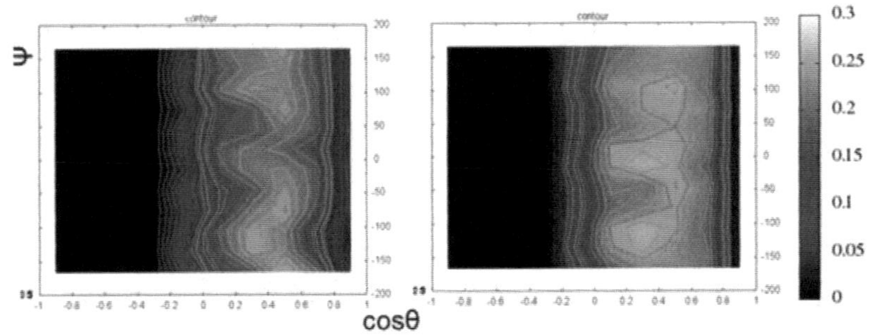

Figure 5. The cosθ-Ψ graph employing data from area (1). MD (left), RMC (right).

The cosθ-Ψ graphs (figure 5) do not display any clear probable orientations. Since in this case ψ determines a rotation around the antiparallel C-Cl bond, any orientation set around this axis is possible. However, three slightly discernible areas on these graphs, due to interatomic repulsion, favor a face to face inverted configuration such as that showed in figure 6.

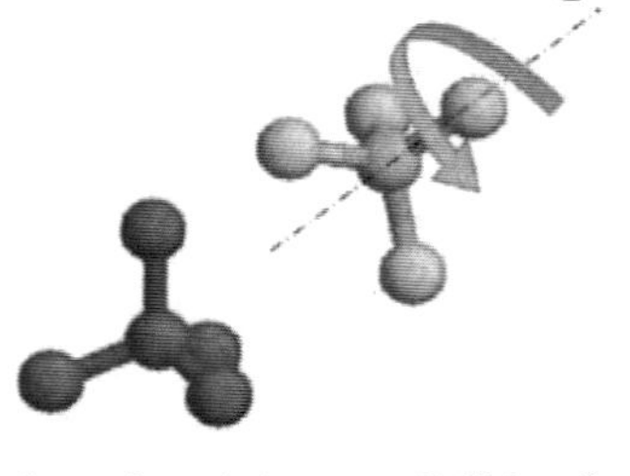

Figure 6. Face to face orientational configuration. Axis set at C-Cl bond perpendicular and opposite to a face of the central molecule.

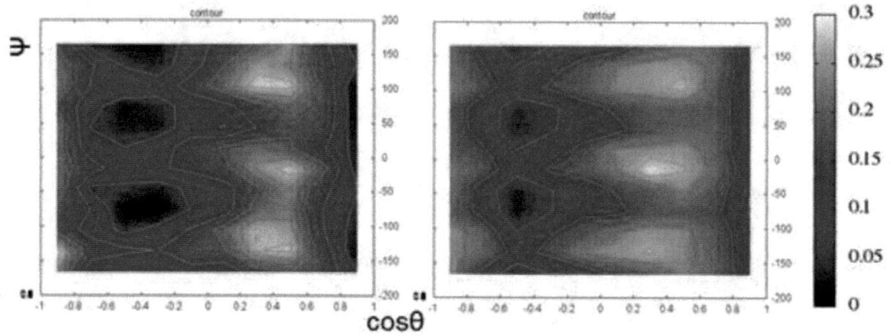

Figure 7. The cosθ-Ψ graph using data from area (2). MD (left), RMC (right)

Now if only the information from area (2) is taken into account, three main orientations can be discerned from the graph (figure 7). These correspond to three equilibrium orientations set 120° apart around an axis directed along any of the three remaining C-Cl bonds; due to the tetrahedral symmetry of the molecule, each of these orientations are essentially the same.

4. Conclusions

The short range order has been successfully determined assessing the six degrees of freedom, three positional and three orientational, describing the molecular order within liquid carbon tetrachloride. The obtained structure is coherent with the geometry of the molecule, as each molecule is set on the faces of the tetrahedral central molecule.

Concerning the relative orientation of two molecules, a face to face configuration is the most probable. Aside from small interferences due to interatomic interaction, the orientation of the molecule around the axis defined by the C-Cl bond perpendicular to a face of the central tetrahedron (Ψ angle) remains unclear. Similar conclusions have been recently obtained for quasi-tetrahedral molecules [5].

Finally, we conclude that for simple tetrahedral molecules both RMC and MD methods provide accurate and similar results regarding short range order. RMC, nevertheless, manages a better fit for experimental data.

5. Appendix

The Euler angles denote a series of rotations around the axes of a coordinate system, in order to describe an orientation. The angles are calculated with the following formulae (ZY'Z'' convention about mobile frame axes):

$$\varphi = \frac{z_y}{|z_y|} \arccos \frac{z_x}{\sqrt{1-z_z^2}} \quad (1) \qquad \theta = z_z \quad (2) \qquad \psi = -\frac{y_z}{|y_z|} \arccos \frac{-x_z}{\sqrt{1-z_z^2}} \quad (3)$$

References

[1] Allen M P and Tildesley D 1987 *Computer simulation of liquids* (Clarendon Press)
[2] Evrard G and Pusztai L 2005 *J. Phys.: Condens. Matter* **17** S1; Gereben O, Jóvári P, Temleitner L and Pusztai L 2007 *J. Optoelectron. Adv. M.* **9**
[3] Narten A H, Danford M D and Levy H A 1967 *J. Chem. Phys.* **46** 4875
[4] Egelstaff P A, Page D I and Powles J G 1971 *Mol. Phys.* **20** 881
[5] Jóvári P, Mészáros G, Pusztai L and Sváb E 2001 *J. Chem. Phys.* **114** 8082
[6] Rey R 2007 *J. Chem. Phys.* **126** 164506
[7] Pothoczki Sz, Temleitner L, Jóvári P, Kohara S and Pusztai L 2009 *J. Chem. Phys.* **130** 064503
[8] Jedlovszky P 1997 *J. Chem. Phys.* **107** 7433
[9] Pardo L C, Tamarit J Ll, Veglio N, Bermejo F J and Cuello G J 2007 *Phys. Rev. B* **76** 134203
[10] Pothoczki Sz, Ottochian A, Rovira-Esteva M, Pardo L C, Tamarit J Ll and Cuello G J 2012 *Phys. Rev. B* **85** 014202

6th Meeting of the Spanish Neutron Scattering Association (SETN2012) IOP Publishing
Journal of Physics: Conference Series **549** (2014) 012015 doi:10.1088/1742-6596/549/1/012015

Structure factor of liquid *n*-butanol at room temperature

G. J. Cuello[1], **V. Cristiglio**[1], **M. A. González**[1]
[1] *Institut Laue Langevin, 6 rue Jules Horowitz, F-38042 Grenoble Cedex 9, France*

C. Cabrillo[2]
[2]*Instituto de Estructura de la Materia, CSIC, Serrano 123, 28006 Madrid, Spain*

Abstract. The structure factor of deuterated liquid n-butanol ($C_4D_{10}O$) has been investigated by neutron diffraction with the aim of exploring the physical nature of the observed pre-peak. The experiment was conducted using two neutron diffractometers: D16 to get a detailed structure factor in the low-Q range, and D4 for a high precision structure factor and proper normalization. In this way a total structure factor was determined covering an extended Q-range from 0.04 to 23.4 Å^{-1}. A molecular dynamics simulation using the general all-atom ab initio force field COMPASS was also carried out. The agreement between experimental and simulated data is very good, giving a plausible interpretation of the origin of the pre-peak observed at 0.6 Å^{-1} as coming from the intermolecular ordering in the liquid.

1. Introduction

Alcohols are important compounds widely used as solvents and many of their particular properties are due to the existence of intermolecular hydrogen bonds. The structure of lowest members of the series have in the liquid phase been determined by neutron diffraction, such as methanol, ethanol and iso-propanol [1, 2, 3, 4, 5, 6, 7].

The signature of H-bonded molecular association in diffraction data is the formation of a pre-peak at a scattering vector, $Q \approx 0.7$-0.8 Å^{-1}, before the main diffraction peak [8, 9]. The pre-peaks of liquid alcohols which are well known in X-ray diffraction data [10], have been detected in the neutron diffraction pattern only recently [8, 9].

The neutron scattering technique, which is sensitive to the positions of both the C and H nuclei, the latter being almost invisible to X-rays. The extremely high value of the incoherent neutron cross section of hydrogen and its exiguous coherent scattering would make it impossible to extract from the total signal the contribution of the partial correlations. We therefore used deuterated n-butanol exploiting the much more favorable coherent ($\sigma_{\rm coh}$ = 5.592 b, 1 b= 10^{-28} m^2) to incoherent ($\sigma_{\rm inc} = 2.05$ b) cross sections ratio of deuterium.

In our study we have determined the structure factor of liquid n-butanol at room temperature (300 K). Using molecular dynamics (MD) simulation we can get data of intra- and inter-molecular distances, partial coordination number (CN) and molecular conformational distribution. The theoretical formalism for the extraction of the structural information from a neutron diffraction experiment is presented elsewhere [11].

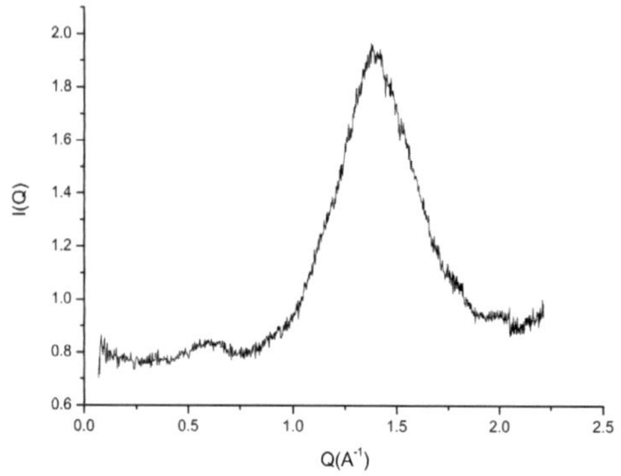

Figure 1. Neutron diffractogram of liquid n-butanol at room temperature as measured in D16. The pre-peak at 0.6 Å is clearly observed.

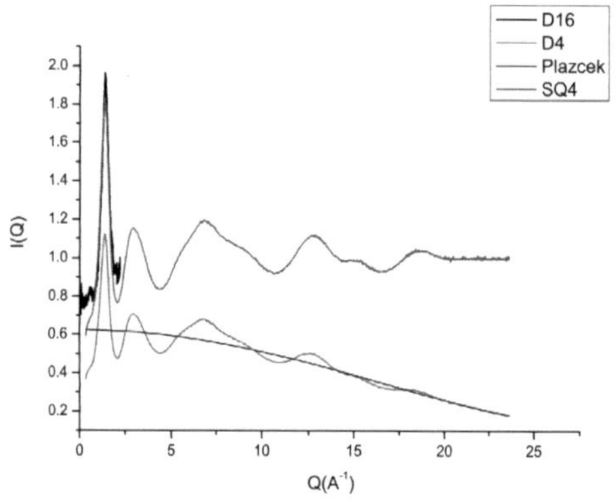

Figure 2. Combined low- and high-Q diffractograms for liquid n-butanol at room temperature as obtained from D4 and D16, after corrections. The corrected D4 (SQ4) data are obtained taking into account the Plazcek contribution.

2. Experimental

A liquid deuterated n-butanol sample of high purity (Aldrich, anhydrous grade, 99.8%) was used for all neutron diffraction experiments. Samples were contained in sealed cylindrical vanadium containers of 6.8/7.0 mm of internal/external diameter. The macroscopic density for liquid deuterated n-butanol at room temperature is 0.92 g/cm^3 or 0.0986 at./Å^3, and its isothermal compressibility is 1.06×10^{-9} Pa^{-9}.

For the high-Q region of the diffractogram the experiment was conducted at the hot neutron diffractometer (D4) [12] optimized for structural investigations of liquids and amorphous. The short-wavelength, 0.49667(1) Å, of incident neutrons allows to cover a Q-range of 0.35-23.5 Å^{-1}. This wavelength as well as the zero angle correction (0.0185°) to the scattering angle were determined using a standard Ni powder sample. In addition, diffraction patterns were measured

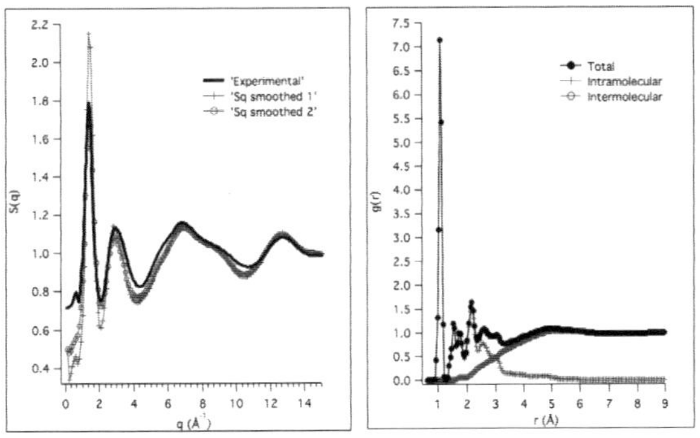

Figure 3. Left: Structure factor as obtained from simulations and its comparison with experimental $S(Q)$. Right: Pair correlation function and the intra- and inter-molecular contributions.

for the sample, the empty sample container, the instrument background and a 6.08 mm diameter vanadium rod, with a beam of 5.0×1.3 mm^2. The low-Q region, 0.01-2.5 Å^{-1} (Fig. 1), was measured on D16 using the same sample and container with an incident wavelength of 4.741 Å. The beam size at the sample position was 11.9×11.7 mm^2 and the experiment was performed at room temperature for the sample, the empty sample container and the instrument background. The differential cross section was obtained after the raw data were corrected for background, absorption, multiple scattering, inelasticity effects and normalization using CORRECT [13].

The MD simulations were carried out in an NVT ensemble consisting of 1000 flexible molecules with density 0.8098 g cm^{-3}, corresponding to a cubic box of side 53.4 Å. The equations of motion were integrated with a time step of 1 fs by the Verlet velocity integrator. Non-bonded interactions were calculated with a cutoff of 12.5 Å for van der Waals forces and using Ewald summation for Coulomb interactions. The equilibrated MD trajectory was recorded for more than 200 ps. The neutron differential cross section was determined using the program package NMOLDYN [14] and the stored MD trajectories obtained using Material Studio.

3. Results

The pre-peak is clearly observed in the low-Q region as measured on D16 (Fig. 1), but on D4 it is more difficult to be observed because its very weak intensity and the low resolution of this instrument. The Fig. 3 shows a preliminary comparison of the structure factors obtained from the numerical simulations and the experimental structure factor, $S(Q)$; in the right frame we observe the comparison of the pair distribution function and the inter- and intra-molecular contributions. The experimental peaks are clearly lower and broader than the simulated ones, because the neighbour distances cannot be resolved and appear as a single peak, contrary to what happens in the simulation. The first and most intense peak of $g(r)$ appears at 1.06 Å and includes the contribution of D–O and C–D correlations. The following peak at 1.57 Å is smaller and wider than the previous and is attributed to four contributions, C–O, C–C, D–D, D–O. The third peak at 2.56 Å is composed by C–D, D–O and D–D (Fig. 3). The positions of the three maxima in X-ray are 0.88, 1.51, 2.5 Å and the first one shows the biggest difference when compared with neutron diffraction [10]. In order to obtain the CN we calculated the total correlation function $T(r)$ by Fourier transformation of the structure factor. The results of Gaussian fits to the peaks are given in Table 1. These data compare very well with X-ray data [10], for which the shortest intra-molecular C–O and C–C distances are 1.41 and 1.52 Å, and the CN are 1.0 and 1.5, respectively.

Table 1. Distances and coordination numbers obtained from the simulated structure.

Bond	Distances (Å)	CN	Bond	Distances (Å)	CN
D–O	0.96	0.96	D–O	1.73	1.08
C–D	1.10	2.26	C–D	2.17	6.18
C–O	1.42	0.99	C–O	2.43	1.0
C–C	1.53	1.48	D–O	2.78	1.52
D–D	1.77	1.15			

The total correlation function shows a first inter-molecular shell at 1.73 Å with a CN of 1.08; a second shell is also present at 3.37 Å with a CN of 0.96. For comparison, the first coordination shell for H-bonds for methanol and ethanol are at distances of 1.75 Å and 2.6 Å, respectively, which confirms the decrease of this distance with the number of carbon atoms in the molecule.

Analyzing the O–O intermolecular distances we observe the first- and second-coordination shells at 2.76 Å and 4.83 Å, with CN 1.54 and 3.07, respectively. Comparing with data from X-ray of O–O (2.79 Å, CN= 1.8) [15] we observe that the distance is the same but the CN decreases. The coordination number of O–O describes intermolecular order and suggests that the cluster structures of these alcohols resemble each other.

4. Conclusions

We have measured the structure factor of liquid n-butanol at room temperature over a wide range of scattering vector (0.01 $Å^{-1} < Q < 23.5$ $Å^{-1}$). By neutron diffraction we can confirm the existence of a pre-peak at 0.6 $Å^{-1}$, which was already observed by X-ray diffraction. Thanks to the good agreement with experimental data, the MD simulations allow us to obtain the partial correlation functions and the correlation distances and CN. Our results are comparable to the previous structure obtained by X-ray diffraction.

The position of the pre-peak in reciprocal space corresponds to a distance of 10.5 Å in real space, which is the inter-molecular distance for molecules aligned along the principal axis. In this way the pre-peak is the signature of H-bonding between molecules. Ongoing research on the temperature dependence of the intensity and width of this pre-peak would help in the interpretation of its physical nature.

References

[1] Benmore C J, Loh Y L 2000 *J. Chem. Phys.* **112** 5877
[2] Yamaguchi T, Benmore C J, Soper A K 2000 *J. Chem. Phys.* **112** 8976
[3] Adya A K, Bianchi L, Wormald C J 2000 *J. Chem. Phys.* **112** 4231
[4] Yamaguchi T, Hidaka K, Soper A K 1999 *Mol. Phys.* **97** 603
[5] Tanaka Y, Ohtomo N, Arakawa K 1985 *Bull. Chem. Soc. Japan* **58** 270
[6] Cuello G J, Talón C, Cabrillo C, Bermejo F J 2002 *Appl. Phys. A: Mat. Sci. Process.* **74** S552
[7] Talón C, Bermejo F J, Cabrillo C, Cuello G J, González M. A, Richardson Jr J W, Criado A, Ramos M A, Vieira S, Cumbrera F L and González L M 2002 *Phys. Rev. Lett.* **88** 115506
[8] Karmakar A K, Krishma P S R and Joader R N 1999 *Phys. Lett. A* **235** 207
[9] Zetterström P, Dahlborg U and Howells W S 1994 *Mol. Phys.* **81** 1187
[10] Vahvaselka K S, Serimma R and Torkkeli M 1995 *J. Appl. Cryst.* **28** 189
[11] Cuello G J 2008 *J. Phys.: Condens. Matter* **20** 244109
[12] Fischer H E, Cuello G J, Palleau P, Feltin D, Barnes A C, Badyal Y S and Simonson J M 2002 *Appl. Phys. A: Mat. Sci. Process.* **74** S160.
[13] Howe M A, McGreevy R L and Zetterström P 1996 Computer code CORRECT, NFL Studsvik
[14] Rog T, Murzyn K, Hinsen K and Kneller G 2003 *J. Comput. Chem.* **24** 657
[15] Shmytko I M, Jiménez-Riobóo R J, Hassaine M and Ramos M A 2010 *J. Phys.: Condens. Matter* **22** 195102

Journal of Physics: Conference Series **549** (2014) 012016 doi:10.1088/1742-6596/549/1/012016

Magnetic moment distribution modeling in non stoichiometric Ni—Mn—Ga ferromagnetic shape memory alloys

P Lázpita[1,a], J Gutiérrez[1], J M Barandiarán[1], V A Chernenko[1,2], C Mondelli[3,4], L. Chapon[3]

[1] BCMaterials and Basque Country University (UPV/EHU), Bilbao, Spain
[2] Ikerbasque, Basque Foundation for Science, Bilbao 48011, Spain
[3]Institut Laue Langevin 6, Rue Jules Horowitz BP 156, 38042 Grenoble Cedex 9, France
[4] CNR-INFM OGG & CRS-SOFT, c/o ILL, 6 Rue Jules Horowitz, BP 156, 38042 Grenoble Cedex 9, France

[a]Email: patricia.lazpita@ehu.es

Abstract. Neutron polarized diffraction technique has been used to elucidate the magnetic moment distribution density in non stoichiometric Ni—Mn—Ga single crystals. These experiments allow us to determine a localized magnetic moment in the Mn position in the austenitic phase, and to validity qualitatively previous models of magnetic distributions where there are antiferromagnetic and ferromagnetic coupling for Mn atoms that are sited out of their properly positions. This measurements show the deep dependence of the magnetic moment with the composition and the atomic order.

1. Introduction

Ferromagnetic shape memory alloys have been focus of interest during the last years due to their unusual magnetomechanical properties. The ferromagnetic thermoelastic martensitic transformation that takes place in the Heusler Ni—Mn—Ga alloys produces a multivariant martensitic phase [1, 2] that make these new multifunctional materials capable of activated twin boundary motion in a controllable way by magnetic field reach strains up to 10 % [3]. This effect is the so called magnetic field induced strain (MFIS) effect [4]. The martensitic structure and the net magnetic moment, key parameters to develop the MFIS effect, have been shown to be extremely sensitive with the composition of these alloys. Moreover they have a great influence on the motion of twin boundaries affecting the output strain. This makes a deep study of the structure and magnetic interactions in such alloys desirable.

Already crystal structures of many off stoichiometric compositions have been studied [5-7]. In previous works Richard et al. [8-10] have been purposed a preferential atomic order for Ni—Mn—Ga alloys with compositions of 44-52 at.% nickel, excess manganese and gallium defective. They present the next atomic distribution: for the alloys that present Ni and Ga defective and excess of Mn, the Mn excess occupies the Ni and Ga vacancies places. Meanwhile, those that present both excess on Ni and Mn, the Ni atoms go to the Mn positions, displacing the excess and the properly placed Mn to the Ga positions. Based on this atomic preferential distribution, and taking into account the general tendency of exchange magnetic coupling that change from ferromagnetic to antiferromagnetic order at reduced interatomic spacing a new simple model of distribution of magnetic moment was proposed [10].

In order to study in a deeper way the magnetic moment distribution in non stoichiometric Ni—Mn—Ga single crystals we have performed a study by neutron polarized diffraction. These experiments will allow us to confirm the validity of our model.

2. Experimental procedure

Single crystal sample of compositions $Ni_{52}Mn_{26}Ga_{22}$, $Ni_{49}Mn_{30}Ga_{21}$ and $Ni_{43}Mn_{38}Ga_{19}$ were grown by the Bridgman technique at the Ames Laboratory (Iowa, USA). For these experiment pieces of approximately 3 mm x 3 mm x 3 mm were used. Martensitic transformation and Curie temperatures were determined by magnetic measurements performed in a Vibrating Sample Magnetometer. The magnetic moment at different temperatures (from 330 K to 400 K) was determined from magnetization curves with applied magnetic fields up to 2 T.

Preliminary experiments were performed to determine the single crystals nuclear structure on the D10 (λ = 2.36Å) and D15 (λ = 0.85 Å and 1.17 Å) instruments at ILL (Grenoble, France). All the single crystals were measured at temperatures in the ferromagnetic and paramagnetic austenitic phase, and so the atomic order could be determined.

Polarized neutron diffraction measurements at different temperatures in the austenitic phase were carried out at D3 instrument at ILL. The crystals were mounting with it [001] axis vertical inside a superconducting magnet. The data collected were treated with the Cambridge Crystallography Subroutine Libraries (CCSL) obtaining the magnetic structure factors. Maximum Entropy method [11] using *MemSys* code available on the ILL was used for direct visualization of the atomic magnetization density maps using the Fourier transformed. Neutron diffraction data analysis was performed using flipping ratios fits with the FullProf software [12].

3. Results and discussion

3.1. Magnetic measurements.

Martensitic transformation (T_M) and Curie (T_C) temperatures have been determined from the low applied magnetic field magnetization curves (Figure 1) and are summarized in Table 1. For all compositions T_M is lower than T_C. This allows us to study both paramagnetic austenitic phase and ferromagnetic austenitic one, and to determine in a better way the magnetic moment distribution in the austenitic ferromagnetic phase. The total macroscopic magnetic moment at different temperatures was determined from hysteresis loops obtained in the austenitic phase (Figure 2). Magnetization curves appear to be saturated for applied magnetic fields of about 0.5 T in the ferromagnetic phase, and they reduce to a clear paramagnetic trend at temperatures higher than the Curie one.

3.2. Neutron diffraction measurements.

Neutron diffraction experiments in the paramagnetic austenite phase were done to confirm the $L2_1$ structure, to determine the cell parameter and the atomic distribution at different position for all studied single crystals. These results were used to analyze the intensities obtained in the polarized neutron diffraction measurements. Figure 3 show the magnetic structure factors at different temperatures in the austenitic phase for applied magnetic field of 2 T. We added the value of the bulk magnetization under 2 T at q=0. As well as the bulk magnetization, the intensity of the magnetic form factor reduces when the temperature is closer to the Curie one. There is no regular decrease of the dependence of the magnetic form factor with the scattering vector which is indicative of the contribution to the total magnetic intensity of Mn and Ni associated to different magnetic form factors. Figure 4 shows the magnetization maps obtained by the maximum entropy

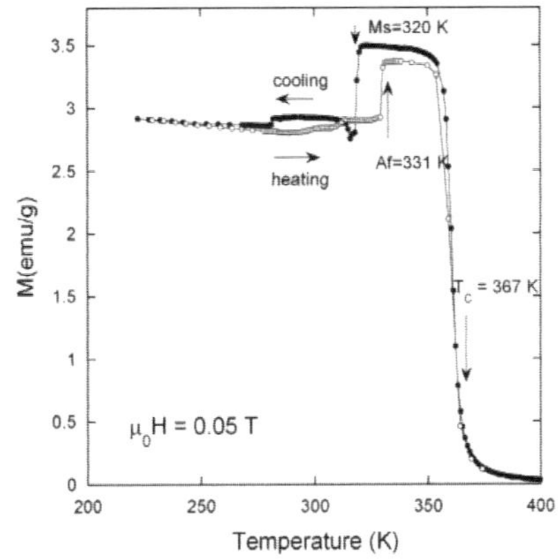

Figure 1. Magnetization of $Ni_{43}Mn_{38}Ga_{19}$ at low magnetic field (0.05 T) as function of temperature.

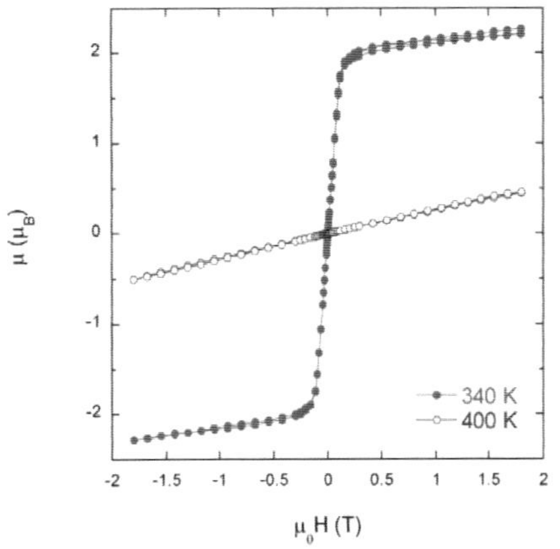

Figure 2. Hysteresis loops of $Ni_{43}Mn_{38}Ga_{19}$ at different temperatures in the austenitic ferromagnetic and paramagnetic phases.

Table 1. Martensitic transformation (T_M) and Curie (T_C) temperatures, cell parameter (a) determined by neutron diffraction, magnetic moment (m) at 2 T determined by magnetic measurements and magnetic moment obtained by neutron polarized diffraction for Mn atoms placed in their own positions (m_{MnMn}), Ni atoms in Ni sites (m_{NiNi}) and Mn atoms placed on Ga positions (m_{MnGa}) respectively, at the austenitic ferromagnetic phase for the studied alloys.

Composition	T_M(K)	T_{Curie} (K)	a (Å)	m (μ_B/Mn)	m_{MnMn} (μ_B)	m_{MnGa} (μ_B)	m_{NiNi} (μ_B)
$Ni_{52}Mn_{26}Ga_{22}$	328	368	5.835	2.01	1.283(8)	-2.6(2)	0.131(5)
$Ni_{49}Mn_{30}Ga_{21}$	298	368	5.813	1.76	1.587(4)	-3.2(1)	0.151(3)
$Ni_{43}Mn_{38}Ga_{19}$	320	367	5.923	1.53	1.814(6)	-1.02(3)	0.098(2)

method. These maps present a mostly localized spin density. Figure 4.a. shows the (x y 0) plane of the $L2_1$ structure. In this plane localized maximums of magnetic density are presented in the positions that correspond to the Mn places verifying that the main contribution to the magnetic moment arise from a localized moment on Mn atoms. On the other hand, the map shows minima at (½ 0 0) positions corresponding to the Ga places. This negative density of the moment matches with the excess of Mn placed at the Ga position (MnGa) that presents an antiferromagnetic coupling with the properly sited Mn (MnMn). Figure 4.b. shows the plane (x y ¼) of the $L2_1$ structure, that contents Ni atoms. A small positive spin density associated with the Ni atoms coupling ferromagnetically with the total magnetic moment.

To determine the magnetic moment distribution at each position we assume an scenario in which all the magnetization is given by the Mn^{2+} and Ni^{2+} spherical free atom form factors site on Ni, Mn or Ga positions depending on the single crystal composition. The average values of magnetic moment determined at different positions are summarized in Table 1. In the stoichiometric Ni_2MnGa alloy with a Curie temperature of 376 K, Brown et al. [13] determined higher values for the magnetic moments of Mn atoms properly sited (2.27 μ_B) at 230 K. This difference can be attributed to the proximity of the Curie temperature in our study and the disorder induced by the

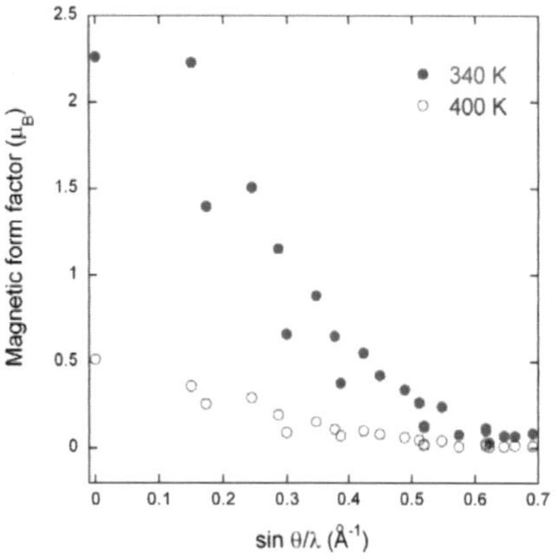

Figure 3. Magnetic structure factors at different temperatures as a function of scattering vector under 2 T.

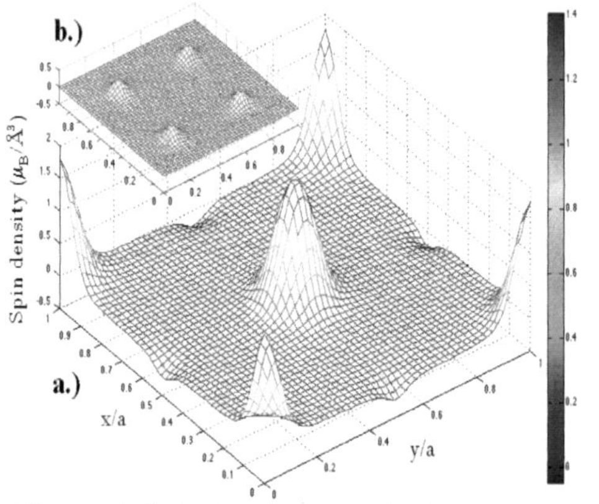

Figure 4. Spin density map of $Ni_{43}Mn_{38}Ga_{19}$ measured at 340 K (2 T) in planes a.) (x y 0) and b.) (x y ¼) of the $L2_1$ unit cell.

non stoichiometric composition. It is remarkable that the magnetic moment average value of the Ni atoms sited at Ni positions (NiNi) for the Ni defective alloy is lower than in the other ones. This fact could be related to the presence of the Mn excess coupling antiferromagnetically with the total moment at the Ni position. Moreover, the Mn magnetic moment behavior depends on the composition and also on the position in the unit cell. In the case of MnMn atoms they show small average values for the Ni excess composition, these results confirm that the excess of Ni occupy the Mn positions prompt the decrease on the total magnetic moment density due the lower magnetic moment linked to the Ni atoms. However, for MnGa atoms the absolute average value is smaller in the lowest Ni content alloy. This can be explain if we take into account the purpose simple model of magnetic moment distribution for this alloy where there is both antiferromagnetic and ferromagnetic coupling of the MnGa atoms. In this case, the drop of the average value of the magnetic moment is related to the sum of the ferromagnetic and antiferromagnetic coupling of the Mn atoms sited at the Ga positions.

4. Conclusions

Polarized neutron diffraction experiments allow us to elucidate about the localized moment nature of the Ni and Mn atoms for non stoichiometric Ni—Mn—Ga single crystals. The determined spin density maps show a negative contribution related to the antiferromagnetic coupling of the Mn sited out of their properly positions. The magnetic moment average value of the Mn depends strongly on the composition and also on their position on the $L2_1$ structure. The low absolute value of the MnGa moment for Ni defective alloy determined is related to the average of the ferromagnetic and antiferromagnetic coupling of the Mn at these positions. With these results we can confirm, at least qualitatively, the validity of the simple model of magnetic moment distribution previously purposed by the authors: MnMn, and NiNi atoms present a ferromagnetic coupling, while MnNi antiferromagnetic one. However, the Mn atoms at Ga positions couple antiferromagnetic or ferromagnetically depending on their nearest neighbors.

Acknowledgements

Authors would like to thank financial support from the Basque Government (ACTIMAT project, ETORTEK programme) and from CICYT of Spain under contract MAT2011-28217-C02-02. P. L. thanks UPV/EHU for a grant. Technical and human support provided by SGiker (UPV/EHU, MICINN, GV/EJ, ESF) is gratefully acknowledged.

Journal of Physics: Conference Series **549** (2014) 012016 doi:10.1088/1742-6596/549/1/012016

References

[1] Webster P J, Ziebeck K R A, Town S L and Peak M S 1984 *Philos. Mag.* B **49** 295-310

[2] Chernenko V A, Cesari E, Kokorin V V and Vitenko I N 1995 *Scr. Metall. Mater.* **33** 1239-1244

[3] Ullakko K, Huang J K, Kantner C, O'Handley R C and Kokorin V V 1996 *Appl. Phys. Lett.* **69** 1966-1968

[4] O'Handley R C, Murray S J, Marioni M, Nembach H and Allen S M 2000 *J. Appl. Phys.* **87** 4712-4717

[5] Brown P J, Crangle J, Kanomata T, Matsumoto M, Neumann K-U, Ouladdiaf B and Ziebeck K R A 2002 *J. Phys.: Condens. Matter.* **14** 10159-10171

[6] Chernenko V A, Cesari E, Pons J and Seguí C 2000 *J. Mater. Res.* **15** 1498-1504

[7] Pons J, Chernenko V A, Santamarta R and Cesari E 2000 *Acta Mater.* **48** 3027-3038

[8] Richard M L, Feuchtwanger J, Allen S M, O'Handley R C, Lázpita P, Barandiarán J M, Gutiérrez J, Ouladdiaf B, Mondelli C, Lograsso T and Schlagel D 2007 *Philos. Mag.* **87** 3437-3447

[9] Lázpita P, Barandiarán J M, Gutiérrez J, Richard M L, Allen S M and O'Handley R C 2008 *Eur. Phys. J. Special Topics* **158** 149-154

[10] Lázpita P, Barandiarán J M, Gutiérrez J, Feuchtwanger J, Chernenko V A and Richard M L 2011 *NJP* **13** 033039 1-14

[11] Papoular R J and Gillon B 1990 *Europhys. Lett.* **13** (5) 429-434

[12] Rodriguez-Carvajal 1993 *Phys. B.* **192** 55-69

[13] Brown P J, Bargawi A Y, Crangle J, Neumann K-U and Ziebeck KRA 1999 *J. Phys.: Condens. Matter* **11** 4715-4722

Journal of Physics: Conference Series **549** (2014) 012017 doi:10.1088/1742-6596/549/1/012017

Study of the transformation sequence on a high temperature martensitic transformation Ni-Mn-Ga-Co shape memory alloy

V. Recarte[1], J. I. Pérez-Landazábal[1], V. Sánchez-Alarcos[1], J.A. Rodríguez-Velamazán[2,3]
[1]Departamento de Física, Universidad Pública de Navarra, Campus de Arrosadía s/n, 31006 Pamplona, Spain.
[2] Instituto de Ciencia de Materiales de Aragón, CSIC - Universidad de Zaragoza, Zaragoza, Spain
[3] Institut Laue-Langevin, CRG's D1B ¬ D15, F-38042 Grenoble, France
E-mail: recarte@unavarra.es

Abstract. Ni-Mn-Ga alloys show the highest magnetic-field-induced strain among ferromagnetic shape memory alloys. A great effort is being done in this alloy system to increase the application temperature range. In this sense, the addition of small amounts of Cobalt to NiMnGa alloys has been proved to increase the MT temperatures through the increase of the electron per atom relation (e/a). In this work, the analysis of the crystal structure of the present phases and the phase transformations has been performed on a Ni-Mn-Ga-Co alloy by neutron diffraction measurements from 10 K to 673 K. The study has been completed by means of calorimetric and magnetic measurements. On cooling the alloy undergoes a martensitic transformation from a face centered cubic structure to a nonmodulated tetragonal martensite. The appearance of intermartensite transformations can be disregarded in the whole temperature range below the martensitic transformation. However, a jump in the unit-cell volume of the tetragonal martensite has been observed at 325 K. Since this temperature is close to the Curie temperature of the alloy both, the structural and magnetic contributions are taken into account to explain the results.

1. Introduction

Ferromagnetic shape memory alloys (FSMA) have attracted great attention as potential actuator materials in recent years [1]. Among these systems, giant magnetic field induced strains of 9.5 % were successfully obtained by rearrangement of martensite variants in ternary Ni-Mn-Ga [2]. In this ternary system a strong dependence of Martensitic Transformation (MT) temperatures [3], crystal structures [4], and magnetocrystalline anisotropy [5] on composition has been observed. The average number of valence electrons per atom (e/a) plays a special role on the stability of martensitic structures and transformation temperatures [6]. Besides, the influence of atomic order on transformation temperatures has been reported in these alloys [7-9]. Therefore the e/a relation and the atomic order are the main factors controlling the transformation temperatures.

In Ni-Mn-Ga-Co alloys with substitution of Mn or Ga by Co the MT temperatures are higher than those in Ni-Mn-Ga alloys [10-13], and the Curie temperature is increased with the substitution of Ni by Co [11-13], so this quaternary alloy is a good candidate for high temperature applications under selected addition of Co.

In this work, the analysis of the crystal structure of the present phases and the phase transformations has been performed on a $Ni_{50}Mn_{26}Ga_{20}Co_4$ alloy by neutron diffraction measurements from 10 K to 673 K. The study has been completed by means of calorimetric and magnetic measurements.

2. Experimental

A polycrystalline ingot of $Ni_{49.95}Mn_{25.9}Ga_{20.2}Co_{3.9}$ alloy was prepared from high purity elements by arc melting under protective Ar atmosphere. The ingot was homogenized in vacuum quartz ampoules at 1273 K during 24 hours. Small samples for calorimetric measurements were obtained from a disc previously cut from the centre of the ingots by slow speed diamond saw. Powder samples for neutron diffraction experiments were elaborated by crushing the samples on an agate mortar. The composition of the samples was analyzed by EDS in a JSM-5610LV Scanning Electron Microscope (SEM). Annealing treatment of 30 minutes at 1173 K followed by quenching into ice water was performed in a vertical induction furnace. Differential Scanning Calorimetry (DSC) measurements in a TA Q100 DSC and thermogravimetric measurements in a TA TGA-29500 were carried out to detect the presence of the different transformations. Powder neutron diffraction measurements were performed at the D1B and D20 installations at the Institute Laue-Langevin. The neutron wavelengths were 1.28 Å and 1.87 Å respectively. The FullProf2000 program [14] was employed to carry out the Rietveld refinement of the different spectra.

3. Results and discussion

Fig. 1 shows the DSC cooling and heating measurements carried out on the as-quenched sample. On cooling an exothermic peak is observed at ≈540 K, linked to the forward MT (austenite→martensite) with the characteristic MT temperatures M_s=573 K (Martensite start), and M_f=485 K (Martensite finish). On the other hand, the reverse MT (martensite→austenite) temperatures were determined from the endothermic peak detected on heating, A_s=523 K (Austenite start), and A_f=615 K (Austenite finish). The MT enthalpy ΔH=8.7±0.2 J g^{-1} was estimated from the average of the cooling and heating values, while the transformation entropy was calculated as $\Delta S=\Delta H/T_o$ =15.8 ±0.2 J kg^{-1}, with T_o equal to average temperature between forward and reverse MT. The thermogram shows a change in the slope of the baseline both on cooling and heating runs at 325 K. Taking into account the characteristics of this change, lack of latent heat and hysteresis between cooling and heating, it must be related to a second order transition as the para-ferromagnetic one. TGA measurements were carried out, see inset in Fig 1, to confirm the presence of this magnetic transition.

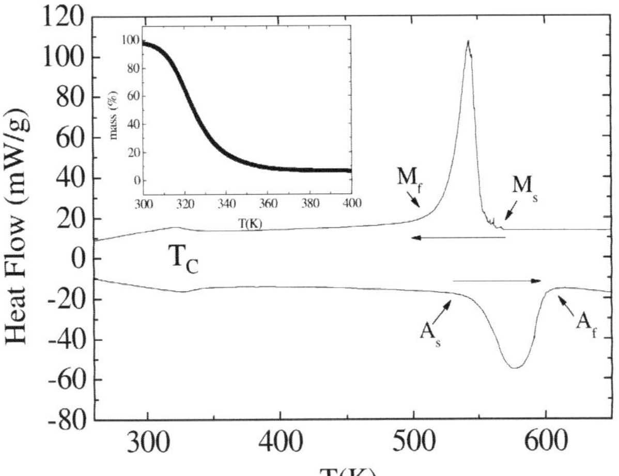

Figure 1. DSC cooling-heating thermogram. Inset, TGA measurement.

The transformation sequence on cooling is the following: a MT from a paramagnetic austenite to a paramagnetic martensite occurs at high temperatures followed by a magnetic ordering of the martensite phase at lower temperatures.
In order to determine the structure of the phases, their thermal evolution and the possibility of intermartensitic transitions neutron diffraction measurements were carried out in the whole temperature range between 10 K and 673 K. Fig, 2-a and 2-b show the neutron diffraction pattern of the austenite at 548 K and the martensite at 448 K respectively with the corresponding Rietveld refinement.

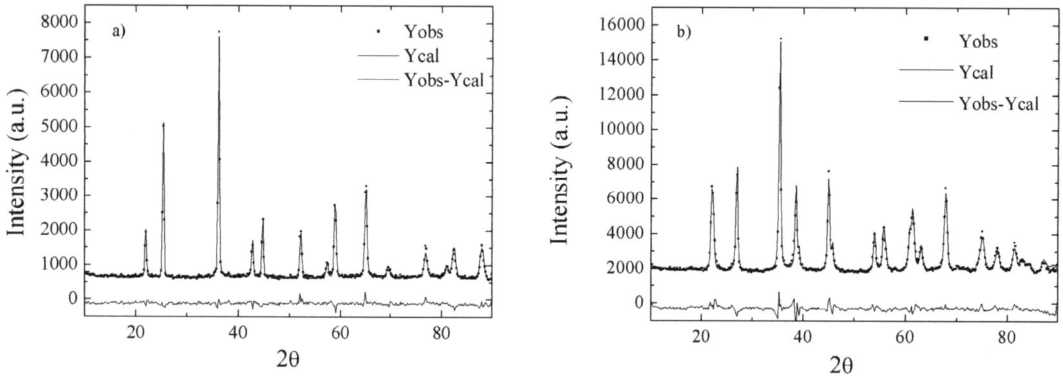

Figure 2. Rietveld refinement of the neutron diffraction patterns obtained at (a) 548 K, austenite phase and (b) 448 K, martensite phase.

The austenite phase has the same crystallographic structure $L2_1$ (space group $Fm\bar{3}m$) than the ternary stoichiometric Ni_2MnGa, where (0,0,0) positions are occupied by Mn atoms, (1/2, 1/2, 1/2) positions by Ga atoms and (1/4, 1/4, 1/4) positions by Ni atoms. According to [15], the spectra shown in Fig. 2a was refined with the Co atoms at the (0,0,0)-Mn positions and with the remaining Mn atoms at the (1/2, 1/2, 1/2)-Ga positions. On the other side, the profile of the martensite structure, as shown in Fig. 2-b, corresponds to a non-modulated (NM) tetragonal martensite (I4/mmm point group) [16] with Mn and Co atoms at the (0, 0, 1/2) positions, Mn and Ga atoms at (0,0,0) and Ni atoms at the (0, 1/2, 1/4) positions. The set of measurements on the martensite phase shows the existence of the tetragonal non-modulated structure in the whole temperature range. No indication of modulated structures is observed because there are no satellites around the Bragg peaks in all the diffraction patterns. Since the NM is the most stable phase of the martensite structures, it can exist at very low temperatures as the last product of intermartensitic reactions. Besides, the NM phase is the only martensite that transforms at high temperatures in alloys with MT temperatures above T_c [6]. The substitution of Mn and Ga by Co has been shown to increase the electron-atom e/a of the alloy, thus increasing the MT temperature and promoting the presence of the NM phase. The calculated electron-atom ratio e/a=7.77 is in agreement with the fact that non-modulated tetragonal martensites (NM) are found for alloys with $e/a \geq 7.7$ in the Ni-Mn-Ga ternary system [6]. So the stability of the martensitic structures and the transformation temperatures of the quaternary alloys with low Co content can be explained in terms of the e/a ratio in the frame of the ternary Ni-Mn-Ga system [17,18].

Both martensite cell parameters show an opposite behaviour as a function of temperature. A negative and positive thermal expansion is observed along c and a axis, respectively (Fig. 3) while the combination of both behaviours induces an overall positive volume thermal expansion, Fig. 4. A similar behaviour has been reported in different materials [19]. Two possible explanations have been suggested. One is due to the Poisson contraction in a direction arising from the large coefficient of expansion in a perpendicular direction [20], and the other is due to changes in Jahn-Teller distortion with temperature [21]. In our case the absolute values of both coefficients of expansion are similar, so

the second explanation seems to be more appropriate, even more taking into account that the martensitic transformation in Ni-Mn-Ga has a band Jahn-Teller origin [22]. On the other hand, the c/a ratio decreases as the temperature approaches to the MT temperature. The evolution of c/a ratio with temperature can be explained taking into account that the c/a ratio converges to the value c/a=1 when the temperature approaches to the MT temperature to minimize the distortion linked to this transition.
In addition, both cell parameters show a slope change at 325 K. The combined changes in a and c cell parameters with temperature induces a jump in the unit-cell volume at the same temperature, Fig.4. The martensite orders magnetically at T_C=325K as DSC and TGA measurements pointed out in Fig. 1. So the magnetic ordering promotes a magneto-volume anomaly reflected as a jump in the unit-cell volume even though the crystallographic structure of the martensite phase does not change in the whole temperature range. The observed lattice contraction with respect to the paramagnetic structure could be related to the magnetoelastic coupling, although no negative thermal expansion [23] has been observed in the present results.

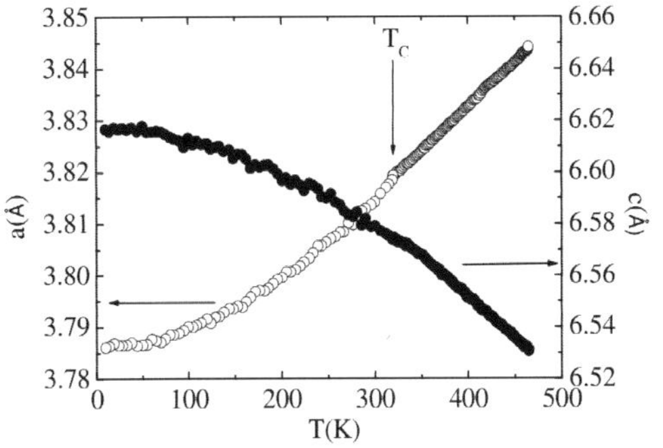

Figure 3. Temperature dependence of the cell parameters of the martensite

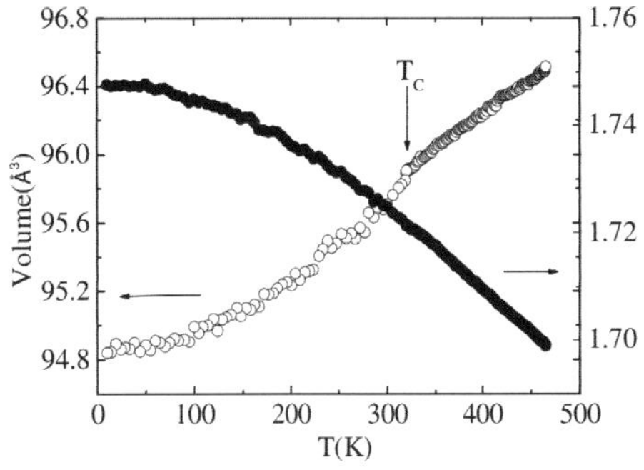

Figure 4. Temperature dependence of the unit-cell volume and c/a ratio of the martensite

Summary and conclusions

The crystal structures and the sequence of the phase transformations have been performed on a $Ni_{50}Mn_{26}Ga_{20}Co_4$ alloy by neutron diffraction measurements from 10 K to 673 K. On cooling the alloy undergoes the martensitic transformation from a cubic $L2_1$ phase to a non-modulated tetragonal martensite. The appearance of intermartensitic transformations can be disregarded in the whole temperature range below the martensitic transformation. However, a magneto-volume effect has been detected linked to the magnetic ordering of the martensite.

Acknowledgements

This work has been carried out with the financial support of the Spanish "Ministerio de Economía y Competitividad" (Projects number MAT2009-07928 and MAT2012-37923-C02-01). The Institute Laue-Langevin, D1B and D20 installations, (Grenoble, France) is acknowledged for the allocated neutron beamtime (Exp. 5-24-261).

References

[1] Kakeshita T *et al.* 2002 *MRS Bull.* **27** 105
[2] Sozinov A *et al.* 2002 *Appl. Phys. Lett.* **80** 1746
[3] Chernenko VA *et al.* 1995 *Scripta Metall.* **33** 1239
[4] Lanska N *et al.* 2004 *J. Appl. Phys.* **95** 8074
[5] Sozinov A *et al.* 2001 *Proceedings SPIE* **4333** 189
[6] Chernenko VA 1999 *Scripta Mater.* **40** 523
[7] Sánchez-Alarcos V *et al.* 2007 *Acta Mater.* **55** 3883
[8] Sánchez-Alarcos V *et al.* 2010 *J. Phys.: Condens. Matter* **22** 166001
[9] Sánchez-Alarcos V *et al.* 2011*Mater. Sci. Forum* **684** 85
[10] Nakamura H *et al.* 2001 *Trans. Mater. Res. Soc. Jpn.* **26** 287
[11] Khovailo V V *et al.* 2003 *Mater. Trans.* **44** 2509
[12] Khan M *et al.* 2005 *J. Appl. Phys.* **97** 10M3004
[13] Cong D Y *et al.* 2008 *Mater. Sci. Eng. A* **473** 213
[14] Rodríguez-Carvajal J 1993 *Physica B* **192** 55
[15] Ayila S K *et al.* 2012 *Phys. Status Solidi* B **249** 620
[16] Wedel B *et al.* 1999 *J. Alloys and Comps.* **290** 137
[17] Sánchez-Alarcos V *et al.* 2008 *Mater. Sci. Eng. A* **481–2** 293
[18] Sánchez-Alarcos V *et al.* 2008 *Acta Mater.* **56** 5370
[15] Krishna Rao K V and Iyengar L 1969 *Acta Crystallogr. A* **25** 302
[20] Krishna Rao K *et al.* 1968 *J. Phys. Chem. Solids* **29** 254
[21] Siratori K and Tida S 1962 *J. Phys. Soc. Jpn.* **17** 208
[22] Brown P J 1999 *J. Phys.: Condens. Matter* **11** 4715
[23] Alvarez-Alonso P *et al.* 2012 *Phys. Rev. B* **86** 184411

Investigating the crystallization process in a FeCrB metallic glass by combining magnetic and neutron thermo-diffraction experiments

Pedro Gorria[1,*], Alejandro Fernández-Martínez[2], J. D. Santos[3], María J. Pérez[3], and Gabriel J. Cuello[4].

[1]Departamento de Física, EPI, Universidad de Oviedo, 33203 Gijón, Spain.
[2]Institut des Sciences de la Terre, Université Joseph-Fourier Grenoble I & CNRS, B.P. 53 38041 Grenoble Cedex 9, France.
[3]Departamento de Física, Universidad de Oviedo, Calvo Sotelo, s/n, 33007 Oviedo, Spain.
[4]Institut Laue Langevin, 6 rue Jules Horowitz, 38042 Grenoble, France

*pgorria@uniovi.es

Abstract. The structural evolution of a $Fe_{70}Cr_{10}B_{20}$ metallic glass was followed by means of *in situ* high-temperature neutron thermo-diffraction and magnetization measurements. Above 723 K the crystallization of bcc-Fe together with a metastable $(FeCr)_3B$ phase with tetragonal crystal structure (space group $I\bar{4}$) is observed. Further heating gives rise to the transformation of the $(FeCr)_3B$ phase into another tetragonal $(FeCr)_2B$ phase (space group $I4/mcm$) + bcc-Fe. On cooling down to room temperature no additional structural transformations occurred. This two-step crystallization process allows understanding quantitatively the intricate variation of the magnetization at high-temperature

1. Introduction

Fe-B based metallic glasses have been widely investigated, due to their technological interest derived from the excellent magnetic and mechanical properties [1], as well as from a basic point of view coming from the relationship between magnetism and the absence of long-range crystalline order [2]. Moreover, it is well known that adding small amounts of transition metal elements, such as Nb, Zr, Mo and others, to the Fe-B metallic glasses gives rise to the formation of a nanocrystalline state after adequate heat treatments, thus improving the soft magnetic response of these alloys [1,3,4]. On the other hand, chromium is usually added to the Fe-B based alloys to enhance the corrosion resistance [5], although it produces a decrease in the values of both the Curie temperature, T_C, and saturation magnetization, M_S [6], in the same way as it occurs in FeCr alloys [7]. Moreover, FeCrB metallic glasses exhibit invar-like features [8] similar to those found in FeZrB alloys [9,10]. Therefore, a good knowledge of the crystallization processes of Fe-B based metallic glasses and the crystallization products is of great importance in order to control their magnetic properties [11].

The crystallization process in binary Fe-B metallic glasses usually takes place in two steps [12-14]. The first step begins when the sample is heated above 650 K, the initial amorphous phase transforms into two crystalline phases, α-Fe with body centered cubic crystal structure (BCC) and $Im\bar{3}m$

symmetry, and metastable Fe_3B with $I\bar{4}$ tetragonal symmetry. The latter is a crystalline phase that only can be stabilized at RT by crystallization of FeB metallic glasses. Heating up at higher temperatures gives rise to a polymorphic transformation which consists in a segregation of Fe atoms from the Fe_3B metastable phase, thus increasing the amount of α-Fe phase, and the transformation of Fe_3B into Fe_2B, also with tetragonal crystalline structure but with higher symmetry ($I4/mcm$). Hence, the final crystallization products are α-Fe and Fe_2B with relative percentages depending on the initial composition of the amorphous as-cast sample.

In this article we describe the connection between the change in the magnetic properties and the crystallization kinetics of a $Fe_{70}Cr_{10}B_{20}$ metallic glass, through a combined study using *in situ* neutron thermo-diffraction experiments and thermo-magnetization measurements. The identification of the crystalline stable and metastable Fe(Cr)B phases appearing along the whole process provides an explanation for the irreversibility of the magnetization vs. temperature curves on heating-cooling cycles.

2. Experimental

Amorphous ribbons with $Fe_{70}Cr_{10}B_{20}$ nominal composition and 5 mm × 25 μm cross section were fabricated by means of a melt spinning setup in a vacuum chamber and using a copper wheel at a speed of 25 ms^{-1}. The amorphous state of the as-cast ribbons was checked by room temperature x-ray powder diffraction (λ = 1.542 Å). The crystallization kinetics of the amorphous ribbons was followed by *in situ* neutron thermo-diffraction at the high-flux D20 two-axis diffractometer (Institute Laue-Langevin, Grenoble, France). The ribbons were cut into 2 cm-long strips (total mass around 200 mg) and introduced into a cylindrical vanadium sample holder for the neutron experiments. Diffraction patterns were collected each 5 min during controlled heating-cooling cycles between 300 and 1073 K at constant rate of 1 K/min, more details concerning similar experiments can be found elsewhere [15,16]. The neutron wavelength was λ = 1.3 Å and the angular range in 2θ was 25-100°. The refinement of the patterns was performed using the Fullprof package [17] based on the Rietveld method. The peak-profile analysis of the diffraction patterns corresponding to the sample already crystallized does not show any broadening originated from small crystals and/or microstrain effects [18]. The magnetization vs. temperature, $M(T)$, curves between 293 and 1073 K were measured in a Faraday susceptometer under an applied magnetic field of 1 kOe. Room temperature hysteresis loops of as-quenched ribbon and several ribbons previously heated up to some selected temperatures were recorded in a conventional homemade induction setup applying a triangular-wave drive field of $H = \pm$ 200 Oe.

3. Results and Discussion

The 3D image and the 2D contour plot of Fig. 1 show the temperature evolution of the neutron powder diffraction patterns between 650 and 1073 K. Below 723 K the diffraction patterns only show broad intensity haloes, thus confirming the amorphous state of the ribbon below such temperature. On heating up above 723 K several intensity peaks begin to grow superimposed to the amorphous contribution, clearly indicating that the crystallization process starts. These peaks can be indexed as the Bragg reflections belonging to α-Fe and $(FeCr)_3B$ phases with body centred cubic (space group $Im\bar{3}m$, a = 2.88±0.01 Å) and tetragonal (space group $I\bar{4}$, a = 8.65 ± 0.01 Å and c = 4.29 ± 0.01 Å) crystalline structures, respectively. This primary crystallization is similar to that occurring in binary Fe-rich FeB metallic glasses, where α-Fe plus metastable Fe_3B are formed [17]. Further heating above 920 K shows that several peaks disappear and new ones emerge (see Fig. 1), denoting that a secondary polymorphic transformation involving the remaining amorphous phase and the $(FeCr)_3B$ crystalline phase takes place. The $(FeCr)_3B$ phase transforms into a stable $(FeCr)_2B$ one with higher tetragonal symmetry (space group $I4/mcm$, a = 5.19 ± 0.01 Å and c = 4.23 ± 0.01 Å). Besides, additional α-Fe is formed during this second transformation as it can be clearly observed in Fig. 1, where the intensity of the Bragg peak situated at around 40° in 2θ suddenly increases.

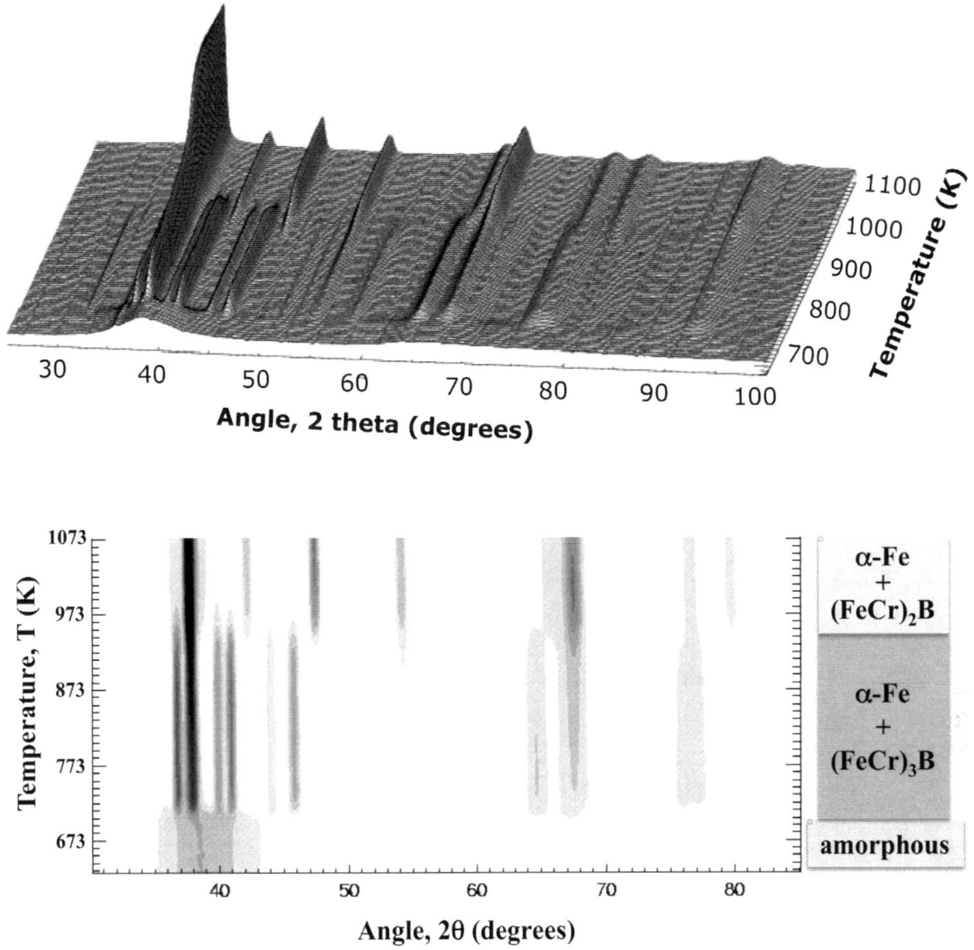

Figure 1. 3-D plot (upper panel) and contour plot (bottom panel) of the temperature evolution of the neutron diffraction patterns obtained with λ = 1.3 Å.

On cooling down to room temperature (the diffraction patterns are not shown) no further structural transformations take place and only Bragg peaks corresponding to α-Fe and $(FeCr)_2B$ are present. The relative amount of each phase determined form the room temperature neutron diffraction pattern is α-Fe (40%) + $(FeCr)_2B$ (60%) [20,21].

In Fig. 2(a) we show the temperature evolution of the magnetization of the as-quenched sample on heating from 290 K up to 1073 K (red full symbols). The $Fe_{70}Cr_{10}B_{20}$ metallic glass is ferromagnetic at room temperature, and the magnetization drops to almost vanishing values for temperatures above the Curie temperature ($T_C \approx$ 400 K). Heating above 723 K up to around 760 K produces a drastic increase of the magnetization, thus suggesting that the crystallization of a ferromagnetic phase with high T_C starts within this narrow temperature interval (the latter can also be observed in Fig. 1, the crystalline peaks growth quickly and their intensities remain almost constant for more than 150 K). Analogous behavior of the $M(T)$ curve has been observed in other FeCrB metallic glasses with similar composition [22].

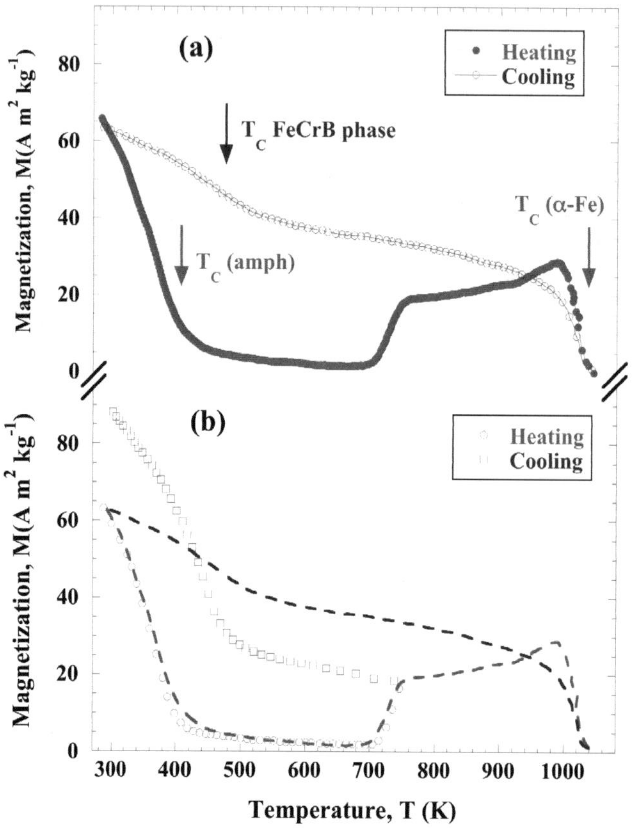

Figure 2. Magnetization vs. temperature, $M(T)$ curves of: (a) the as-quenched sample on heating from room temperature (red full symbols) and the fully crystallized sample on cooling down from 1073 K (blue open symbols); (b) same as (a) but the final temperature is 750 K which corresponds with the end of the first crystallization step [$M(T)$ evolution in (a) is also plotted in (b) as discontinuous lines for comparison, see text for details].

The slight increase of the magnetization between 760 and 900 K could be due to some temperature-induced homogenization of the α-Fe phase (grain growth and elimination of grain boundaries), because the neutron diffraction patterns do not undertake any noticeable change as previously mentioned. However, for $T > 900$ K a second magnetization climb takes place -although the slope is smaller than the previous one- due to the increase in the amount of α-Fe as a consequence of the second eutectic crystallization process (see Fig. 1). Further heating up to $T = 1073$ K only shows a magnetization drop to zero due to the T_C of α-Fe (1044 K). The subsequent cooling down to room temperature shows a completely different $M(T)$ curve (see blue open symbols in Fig. 2a) compared with that measured upon heating the as-quenched sample. The two changes in the slope of the $M(T)$ curve can be ascribed to the Curie temperatures of α-Fe (≈ 1044 K) and the $(FeCr)_2B$ (≈ 500 K).

Moreover, we have measured the $M(T)$ curve by heating the as-quenched sample up to a temperature at which the first crystallization process is finished ($T = 760$ K), and then cool it down to room temperature. It can be observed in Fig. 2b that the although the $M(T)$ curve matches, as expected, with that of Fig. 2a on heating (red open symbols), it is different on cooling because the crystalline phases present in the sample and its relative amount is different. From the neutron diffraction pattern collected at 760 K, we have estimated that the relative amount between α-Fe and $(FeCr)_3B$ is around 20 and 80%, respectively.

We show in Fig. 3 the hysteresis loops recorded at room temperature corresponding to the as-quenched sample and several ribbons heated up to selected temperatures. The loops measured for the ribbons heated up to 573, 723 and 773 K (denoted as @573 K, @723 K and @773 K, respectively) are compared with that of the as-quenched sample in Fig. 3a. It is clear that the loop of sample @573 K is almost identical to that of the as-quenched sample because no structural transformation has started yet below such temperature.

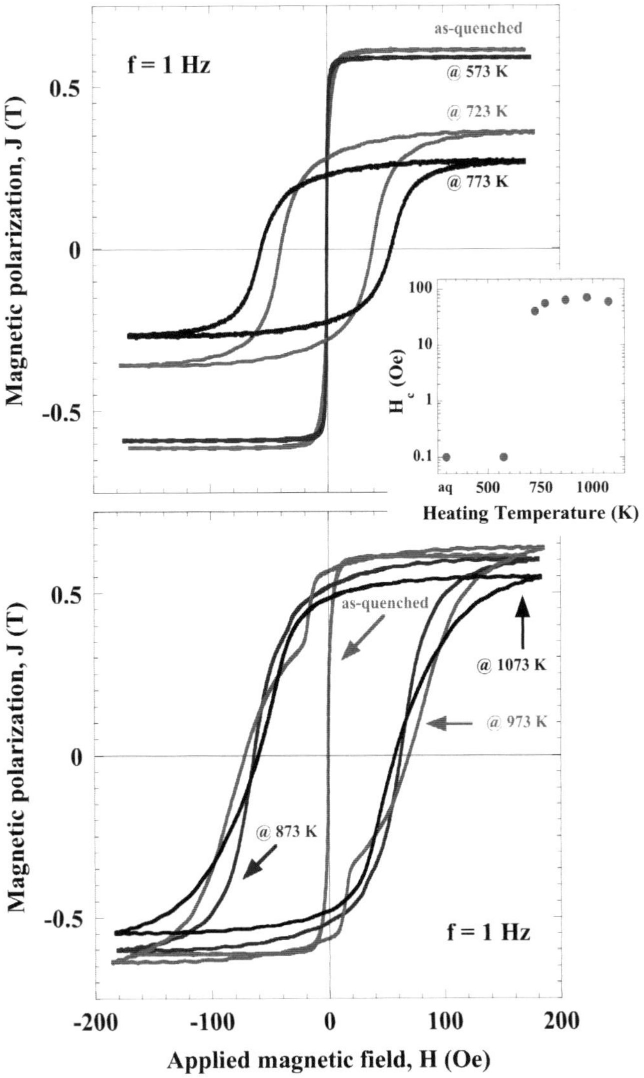

Figure 3. Room temperature hysteresis loops corresponding to the samples as-quenched and heated up to several selected temperatures. The inset shows the evolution of the coercive field.

Samples @723 K and @773 K exhibits a large increase in coercitivity (almost three orders of magnitude) that can be explained on the basis of the crystallization of $(FeCr)_3B$ ferromagnetic phase but with high magnetic anisotropy due to its lower symmetry. Heating at higher temperatures (see Fig. 3b for the hysteresis loops corresponding to the samples heated up to 873, 973 and 1073 K compared that of the as-quenched one) gives rise to an increase in the magnetic polarization because the amount of a-Fe increases, however, as the new $(FeCr)_2B$ formed phase is also highly anisotropic, the coercivity remains very high compared with that of the starting amorphous alloy. The evolution of the coercivity with the temperature of heating is depicted in Fig. 3c.

Conclusions

In summary, we have shown that the $Fe_{70}Cr_{10}B_{20}$ amorphous ribbons undergo a complex crystallization processes on heating from room temperature up to 1073 K, with two differentiated eutectic transformations starting at around 723 K [amorphous → α-Fe + $(FeCr)_3B$ are formed] and 900 K [$(FeCr)_3B$ + remaining amorphous → α-Fe + $(FeCr)_2B$]. Moreover, the combined use of neutron thermo-diffraction and thermo-magnetometry techniques permits the complete understanding of both the irreversible temperature dependence of the magnetization during the heating-cooling processes, and the great increase of around three orders of magnitude in the coercive field values due to the formation of tetragonal FeCrB phases with large magnetic anisotropy. These findings could open new perspectives in the study of the crystallization routes followed by Fe-based metallic glasses with potential interest in technological applications.

Acknowledgments

We thank the Spanish MICINN and FEDER program for financial support under project MAT2011-27573-C04-02. We also thank ILL and Spanish CRG for the allocation of neutron beamtime.

References

[1] McHenry M E *et al.* 1999 *Prog. Mater. Sci.* **44** 291.
[2] *Amorphous Metallic Alloys* 1983, ed F E Luborsky (Butheworths, London, UK).
[3] Gorria P *et al.* 1993 *J. Appl. Phys.* **73** 6600.
[4] Suzuki K and Kadogan J M 2000, *J. Appl. Phys.* **87** 7097.
[5] Zhang L and Manthiram A 1997 *J. Magn. Magn. Mater.* **168** 85.
[6] Güntzel U and Westerholt K 1990 *Phys. Rev. B* **41** 740.
[7] Martínez-Blanco D *et al.* 2011 *J. Alloys Compd.* **509S** S397.
[8] Fukamichi K *et al.* 1982 *J. Appl. Phys.* **53** 2246.
[9] Barandiarán J M *et al.* 1996 *Phys. Rev. B* **54** 3026.
[10] Martínez A *et al.* 2010 *Phys. Rev. B* **82** 012406.
[11] Gorria P *et al.* 1996 *J. Phys: Condens. Matter.* **8** 5925.
[12] Battezzati L *et al.* 1997 *J. Alloys Compd.* **247** 164.
[13] Duhaj P and Švec P 1997 *Mater. Sci. Eng. A* **226-228** 245
[14] Palumbo M *et al.* 2003 *Intermetallics* **11** 1293.
[15] Fernández Barquín L *et al.* 1998 *J. Phys: Condens. Matter* **10** 5047.
[16] Gorria P *et al.* 2009 *Phys. Status Solidi (RRL)* **3** 28.
[17] Rodríguez Carvajal J 1993 *Physica B* **192** 55.
[18] Martinez-Blanco D *et al.* 2008 *J. Phys: Condens. Matter* **20** 335213.
[19] Günterhodt H J *et al.* 1981 *Glassy Metals I, Topics in Applied Physics* vol 46, (Springer, Heidelberg, Germany).
[20] Fernández-Martínez A *et al.* 2005 *J. Magn. Magn. Mater.* **294** e155.
[21] Fernández-Martínez A *et al.* 2007 *J. Non-Cryst. Solids* **353** 855.
[22] Sort J *et al.* 2010 *Small* **6** 1543.

Optimizing the Curie temperature of pseudo-binary $R_xR'_{2-x}Fe_{17}$ (R, R' = rare earth) for magnetic refrigeration

Pablo Álvarez-Alonso[1,*], Pedro Gorria[2], Gabriel Cuello[3], Inés Puente Orench[3,4], José L. Sánchez Llamazares[5], Victorino Franco[6], Marian Reiffers[7] and Jesús A. Blanco[8].

[1]Departamento de Electricidad y Electrónica, UPV/EHU, B° Sarriena, s/n, 48940 Leioa, Spain.
[2]Departamento de Física, EPI, Universidad de Oviedo, 33203 Gijón, Spain.
[3]Institute Laue Langevin, 6 rue Jules Horowitz, 38042 Grenoble, France
[4]Instituto de Ciencia de Materiales de Aragón, CSIC–Univ. Zaragoza, 50009 Zaragoza, Spain.
[5]División de Materiales Avanzados, IPCyT, 78216, San Luis Potosí, Mexico.
[6]Departamento de Física de la Materia Condensada, ICMSE-CSIC, Universidad de Sevilla, P.O. Box 1065, 41080 Sevilla, Spain.
[7]Faculty of Humanities and Natural Sciences, Presov University, ul. 17. novembra 1, SK-08116 Presov, Slovakia
[8]Departamento de Física, Universidad de Oviedo, Calvo Sotelo, s/n, 33007 Oviedo, Spain.

*pablo.alvarez@ehu.es

Abstract. Several pseudo-binary $R_xR'_{2-x}Fe_{17}$ alloys (with R = Y, Ce, Pr, Gd and Dy) were synthesized with rhombohedral Th_2Zn_{17}-type crystal structure determined from x-ray and neutron powder diffraction. The choice of compositions was done with the aim of tuning the Curie temperature (T_C) in the 270 ± 20 K temperature range, in order to obtain the maximum magneto-caloric effect around room temperature. The investigated compounds exhibit broad isothermal magnetic entropy changes, $\Delta S_M(T)$, with moderate values of the refrigerant capacity, even though the values of ΔS_M^{Peak} are relatively low compared with those of the R_2Fe_{17} compounds with R = Pr or Nd. The reduction on the ΔS_M^{Peak} is explained in terms of the diminution in the saturation magnetization value. Furthermore, the $\Delta S_M(T)$ curves exhibit a similar caret-like behavior, suggesting that the magneto-caloric effect is mainly governed by the Fe-sublattice. A single master curve for $\Delta S_M/\Delta S_M^{Peak}(T)$ under different values of the magnetic field change are obtained for each compound by rescaling of the temperature axis.

1. Introduction

A wide family of magnetic intermetallic compounds arises from alloying rare-earth (R) elements and 3d transition metals (M) [1,2]. Localized magnetism of rare-earth sublattice combined with an itinerant magnetism of the 3d sublattice makes these magnetic intermetallics very attractive from both fundamental Solid State Physics and commercial applications. In the case of 3d transition metals, stable R-M compounds were obtained for M = Mn, Fe, Co or Ni, where the unpaired 3d electrons of the transition metal component give rise to a net magnetic moment in most cases. The magnetic interactions in R-M compounds comprise, therefore, three different types: the R-R interaction, the R-M interaction and the M-M interaction [3]. The R-R interaction is the weakest due to the localized character of the electrons.

The Fe-rich R_2Fe_{17} intermetallics have attracted a renewed interest as promising materials for room temperature magnetic refrigeration technology due to a remarkable magneto-caloric effect. For R = Pr or Nd moderate values for the maximum of the isothermal magnetic entropy change, ($\Delta S_M^{Peak} \approx 6$ J/kg K^{-1} for a change of the applied magnetic field $\mu_0\Delta H = 5$ T), together with broad maxima of the $\Delta S_M(T)$ curves, give rise to a large refrigerant capacity (*RC*) [4-6]. In addition, the coexistence of both direct and inverse magneto-caloric effect has been recently discovered in Er_2Fe_{17} compound, which also exhibits large magneto-volume anomalies [7]. From the structural point of view, these compounds present polymorphism with two different types of crystal structure depending on the rare-earth element. From Ce to Sm the R_2Fe_{17} compounds crystallize in the rhombohedral Th_2Zn_{17}-type ($R\bar{3}m$ space group), with the rare earth occupying a unique crystallographic site, 6c (in Wychoff notation [8]), whereas the Fe atoms occupy four non-equivalent positions: 6c, 9d, 18f and 18h [6,9]. In turns, the crystal structure is hexagonal (Th_2Ni_{17}-type, $P6_3/mmc$ space group) for the alloys with R = Dy to Lu. In the latter case, the rare earth occupies two different sites: 2b and 2d, characterized by a rather similar local atomic arrangement of Fe next-neighbour atoms and by a slight different arrangement of the rare-earth atoms. The Fe atoms also occupy four different crystallographic sites: 4f, 6g, 12j and 12e [1,7,10]. Both crystal structures are found for R = Gd and Tb. Moreover, the mixing of two different rare-earth elements in the form $R_xR'_{2-x}Fe_{17}$ can result in a single-phase compound, depending on the R_2Fe_{17} and R'_2Fe_{17} crystal structures [11,12]. Therefore, the combination of two R elements can facilitate the synthesis of single-phase materials with one of the two possible crystal structures, which is sometimes difficult. In this way, it is also feasible to modify the Curie temperature (T_C) of R_2Fe_{17} intermetallics by mixing different rare earth elements [13,14].

In the present work we report on the synthesis, the crystal structure, the magnetic properties and the magneto-caloric response of five pseudo-binary intermetallic alloys: $Y_{1.2}Ce_{0.8}Fe_{17}$, $Pr_{1.5}Ce_{0.5}Fe_{17}$, $Gd_{1.3}Ce_{0.7}Fe_{17}$, $Dy_{1.15}Ce_{0.85}Fe_{17}$ and $YPrFe_{17}$. Such compositions were selected with two main goals: the possibility of tuning the value of the Curie temperature in the immediacy of room temperature and also, how the changes in the R-sublattice could affect the magneto-caloric effect.

2. Experimental

The starting materials for the preparation of as-cast pellets (Goodfellow, 99.9% pure rare earth elements relative to rare earth content and 99.99% Fe pieces) were mixed in the nominal molar ratio 2:17. The polycrystalline ingots were arc melted under Ar atmosphere. Each specimen was wrapped in tantalum foil and sealed under vacuum in a quartz ampoule, and further annealed at 1263 K for 7 or 12 days to ensure their homogeneity, followed by water-quenching. Energy dispersive x-ray spectroscopy (EDS) confirmed the nominal composition of the alloys.

Crystal structure and lattice parameters were studied by means of both x-ray (XRD) and neutron powder diffraction (ND). XRD studies were performed at room temperature in a high resolution powder diffractometer (Seifert model XRD3000) operating in Bragg-Brentano geometry using Cu $K\alpha$ radiation ($\lambda = 1.542$ Å), with scans in 2θ registered between 2° and 130° with $\Delta 2\theta = 0.025°$ steps and counting times of 20 s per point. ND patterns were collected on the D1B high-flux two-axis powder diffractometer ($\lambda = 2.52$ Å) at $T = 300$ K, at the ILL (Grenoble, France). The full-profile analysis of the diffraction patterns was carried out with the FullProf suite package [15]. Peak broadening contributions originated from small crystals and/or microstrain effects [16] were not observed.

Magnetization measurements were carried out in a Quantum Design SQUID and a Lakeshore VSM model 7407 magnetometers. The temperature dependence of the magnetization, $M(T)$, under a low applied magnetic field ($\mu_0 H = 5$ mT) was measured in order to estimate the T_C value of the samples as the minimum of the dM/dT vs. T curve. Isothermal magnetization vs. applied magnetic field curves, $M(H)$, were measured in the temperature range 90 - 390 K. For each $M(H)$ curve the applied magnetic field was increased from 0 up to 1.5 T. The isothermal magnetic entropy variation, $\Delta S_M(T,H)$, due to a change of the applied magnetic field from an initial value $H = 0$ to a final value H is calculated using the Maxwell relation:

$$\Delta S_M(T,H) = \int_0^H \left(\frac{\partial M(T',H')}{\partial T'} \right)_{T'=T} dH' \qquad (1)$$

3. Results and Discussion

Figure 1 depicts the XRD (a) and the ND (b) pattern of $Gd_{1.3}Ce_{0.7}Fe_{17}$ and $Dy_{1.15}Ce_{0.85}Fe_{17}$ alloys, respectively. The other three samples show similar powder diffraction patterns. The observed intensity peaks of the XRD and ND patterns can be indexed as the Bragg reflections corresponding to the rhombohedral Th_2Zn_{17}-type crystal structure with $R\bar{3}m$ space group (#164). There is not any trace of the hexagonal Th_2Zn_{17}-type phase in the diffraction patterns. The main crystallographic information is given in Table 1. From the fit of the diffraction patterns it was evidenced that the R and R' atoms share the 6c site with the same atomic coordinates. Hence, mixing Ce with Y and Gd, as well as Pr with Y stabilizes the rhombohedral phase versus the hexagonal one for the Y_2Fe_{17}, Gd_2Fe_{17} and Dy_2Fe_{17} binary intermetallic compounds. It should be noticed that the Pr containing compounds have the largest cell volume of the studied pseudobinary compounds, as a consequence of the lanthanide contraction. Additionally, a small amount (< 3%) of α-Fe impurity phase ($Im\bar{3}m$ space group) was found in all the compounds, similarly as that occurring in the rhombohedral R_2Fe_{17} binary compounds. The ratio c/a ~ 1.45 is in good agreement with those reported for the rhombohedral R_2Fe_{17} alloys [6,9].

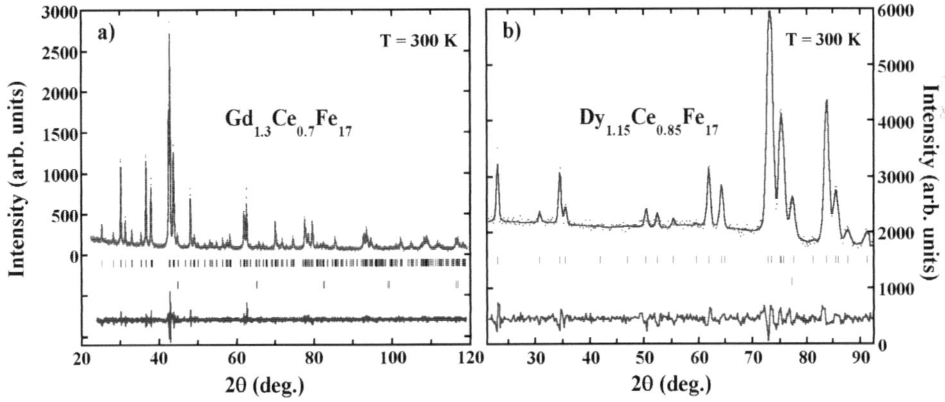

Figure 1. Observed (dots) and calculated (solid line) a) x-ray and b) neutron diffraction patterns collected at $T = 300$ K for $Gd_{1.3}Ce_{0.7}Fe_{17}$ and $Dy_{1.15}Ce_{0.85}Fe_{17}$, respectively. Vertical bars represent the positions of the Bragg reflections; the first row corresponds to the Th_2Zn_{17}-type crystal structure nuclear scattering, whereas the second row corresponds to the α-Fe impurity phase. The observed–calculated difference is depicted at the bottom of each figure.

The Curie temperatures of $Y_{1.2}Ce_{0.8}Fe_{17}$ (253 ± 5 K), $Pr_{1.5}Ce_{0.5}Fe_{17}$ (264 ± 5 K), $Gd_{1.3}Ce_{0.7}Fe_{17}$ (293 ± 5 K), $Dy_{1.15}Ce_{0.85}Fe_{17}$ (273 ± 5 K) and $YPrFe_{17}$ (290 ± 5 K), are in between those reported values for Y_2Fe_{17} (310 ± 4 K) [17], Ce_2Fe_{17} (Néel temperature = 210 K) [18], Pr_2Fe_{17} (286 ± 2 K) [9], Gd_2Fe_{17} (476 K) [19] and Dy_2Fe_{17} (370 K) [20]. Mixing different R ions gives rise to variations in the Fe–Fe next neighbor distances along the c-axis, which strongly affects the magnetic interactions, and thus modifies the value of T_C [7,9]. Therefore, the Curie temperature of the resulting pseudo-binary alloys can be tuned between those values of the pure binary compounds.

The saturation magnetization, M_S, at T = 85 K (see Table 2) was estimated from the fit of the magnetic field dependence of the magnetization to an approach-to-saturation law [21],

$$M = M_S\left(1 - \frac{b}{H^2}\right) + \chi_0 H \quad (2)$$

These results can be understood if it is assumed that the magnetic structure is that of the parent R_2Fe_{17} alloys. The Pr-containing alloys show higher M_S values due to the collinear ferromagnetic structure of Pr_2Fe_{17}. In this sense, $YPrFe_{17}$ and $Pr_{1.5}Ce_{0.5}Fe_{17}$ have similar M_S values, because Y has no magnetic moment and, even though the pure Ce_2Fe_{17} has a complicated magnetic phase diagram, the Cerium ion has a weak magnetic moment [22,23] and its contribution to the total magnetization of the alloy can be neglected. The $Y_{1.2}Ce_{0.8}Fe_{17}$ has an intermediate value because only the Fe-sublattice contributes to the M_S. The antiparallel coupling between the Dy^{+3} and Gd^{3+} magnetic moments and those of the Fe sublattice explains the lower M_S value exhibited by the $Gd_{1.3}Ce_{0.7}Fe_{17}$ and $Dy_{1.15}Ce_{0.85}Fe_{17}$.

TABLE 1. Cell parameters and atomic positions corresponding to the studied compounds (with rhombohedral crystal structure) obtained from room temperature XRD patterns.

	$Y_{1.2}Ce_{0.8}Fe_{17}$	$Pr_{1.5}Ce_{0.5}Fe_{17}$	$Gd_{1.3}Ce_{0.7}Fe_{17}$	$Dy_{1.15}Ce_{0.85}Fe_{17}$	$YPrFe_{17}$
a (Å)	8.490 (1)	8.565 (1)	8.523 (1)	8.487 (1)	8.540 (1)
c (Å)	12.394 (1)	12.487 (2)	12.436 (1)	12.417 (3)	12.419 (1)
c/*a*	1.460	1.455	1.459	1.461	1.454
V(Å^3)	774.1 (1)	790.2 (2)	782.2 (1)	773.1 (2)	784.3 (2)
R-6c: z	0.340 (1)	0.320 (1)	0.344 (1)	0.359 (2)	0.348 (3)
Fe-6c: z	0.098 (2)	0.091 (2)	0.096 (1)	0.125 (6)	0.092 (1)
Fe-18f :x	0.281 (1)	0.287 (3)	0.292 (2)	0.291 (2)	0.293 (1)
Fe-18h: x	0.165 (1)	0.172 (1)	0.168 (1)	0.160 (2)	0.169 (2)
Fe-18h: z	0.481 (2)	0.481 (2)	0.490 (1)	0.482 (2)	0.489 (1)
R_B	11.0	10.8	4.5	5.7	5.7
χ^2(%)	1.5	3.0	1.4	1.2	1.5

According to the saturation magnetization, the Pr-based pseudo-binary intermetallics would exhibit the largest magnetic entropy change because of their higher saturation magnetization values. The isothermal magnetic entropy change, ΔS_M, was calculated from the set of isothermal magnetization vs. applied magnetic field, *M*(*H*), curves. In figure 2 those *M*(*H*) curves corresponding to the $Gd_{1.7}Ce_{0.3}Fe_{17}$ alloy measured between 220 and 390 K with a temperature step of 10 K are shown.

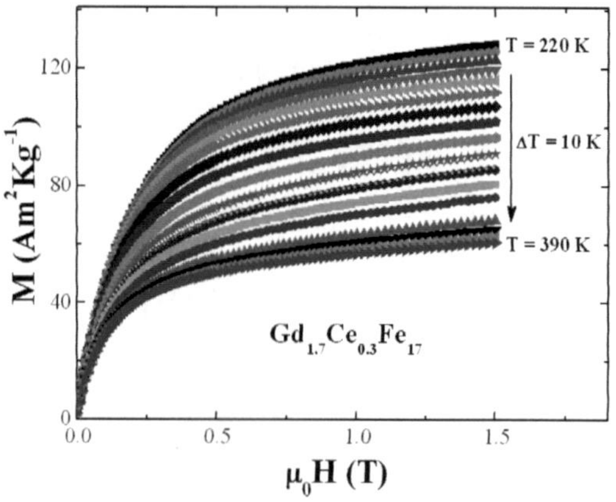

Figure 2. Magnetic field dependence of the isothermal magnetization for $Gd_{1.7}Ce_{0.3}Fe_{17}$.

The Pr-containing pseudo-binary alloys have the largest peak values of the magnetic entropy change (see figure 3), $|\Delta S_M|^{Peak}$, as expected from the relative M_S values. Besides that, the $|\Delta S_M|(T)$ curves are broad, being the full width at half maximum, δT_{FWHM}, over 80 K for $\mu_0\Delta H = 1.5$ T. This is an important parameter to be taken into account when the refrigerant capacity (*RC*) of a material with potential interest in magnetic refrigeration has to be evaluated [5]. This figure of merit can be estimated as the product of the full width at half maximum and the peak value. The field dependence of the refrigerant capacity was also obtained (see inset in figure 3). The values for $Y_{1.2}Ce_{0.8}Fe_{17}$, $Pr_{1.5}Ce_{0.5}Fe_{17}$ and $YPrFe_{17}$ are lower than those of the parent Pr_2Fe_{17} and Y_2Fe_{17} alloys [24], as a direct consequence of the reduction of M_S. However, Ce_2Fe_{17} presents an antiferromagnetic to paramagnetic phase transition, which has low refrigerant capacity [25]. $Dy_{1.15}Ce_{0.85}Fe_{17}$ has increased its *RC* with respect to the pure Dy_2Fe_{17} alloy [25]. As far as we know, no data of Gd_2Fe_{17} was published yet. Even though $|\Delta S_M|^{Peak}$ of the studied compounds is much lower than that of the polycrystalline Gd, the *RC* reaches 65% of RC_{Gd}.

TABLE 2. Magnetic and magneto-caloric properties ($\mu_0\Delta H = 1.5$ T) of the pseudobinary alloys.

	$Y_{1.2}Ce_{0.8}Fe_{17}$	$Pr_{1.5}Ce_{0.5}Fe_{17}$	$Gd_{1.3}Ce_{0.7}Fe_{17}$	$Dy_{1.15}Ce_{0.85}Fe_{17}$	$YPrFe_{17}$	Pure Gd
T_C (K)	253(5)	264(5)	293(5)	273(5)	290(5)	289(5)
M_S (A m^2 kg^{-1})	113	135	87	88	134	-
$\|\Delta S_M^{Peak}\|$ (J K^{-1} kg^{-1})	1.6	2.4	0.8	1.3	2.3	4.3
RC (J kg^{-1})	92	103	83	83	98	160
δT_{FWHM} (K)	57	42	107	65	43	37

It was shown, both theoretically and from a phenomenological approach, that there exists a master curve for the magnetic entropy change in magnetic materials with a second-order phase transition [26-28]. This master curve implies the collapse of the properly rescaled $\Delta S_M(T)$ curves for different values of the applied magnetic field change, and consequently it could be useful to predict the magnetic field and temperature dependences of ΔS_M in those materials which do not follow a mean field approach [29].

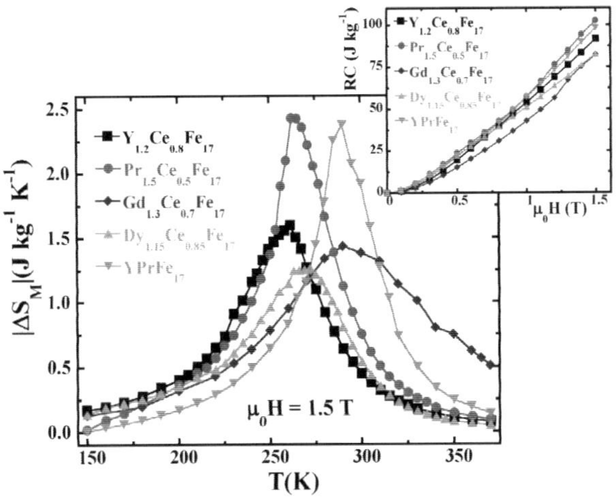

Figure 3. Temperature dependence of the magnetic entropy change for $R_xR'_{2-x}Fe_{17}$ alloys under $\mu_0\Delta H$ = 1.5 T. Inset shows the field dependence of refrigerant capacity.

Therefore, this allows the extrapolation of the $|\Delta S_M(T)|$ curves for different values of the applied magnetic field change, as well as for the estimation of $|\Delta S_M(T)|$ curves in other intermetallic compounds with related compositions, in which the rare earth metal and/or the relative amount is varied. It is theoretically derived from the knowledge of the magnetic equation of state, whereas, in a phenomenological way, it is obtained from the $\Delta S_M(T)$ curves themselves. We will focus in the phenomenological procedure. Firstly, each $\Delta S_M(T)$ curve is normalized to the value of the corresponding ΔS_M^{Peak}. Secondly, the temperature axis is rescaled imposing that a reference point in the $\Delta S_M(T)$ curve corresponding to a previously selected fraction of ΔS_M^{Peak}, makes the rescaled temperature, θ, to be equal 1 ($\theta = -1$) when the temperature at which the reference point, T_r, is chosen is $T_r > T_C$ ($T_r < T_C$) in the following way [30] $\theta = (T - T_C)/(T_r - T_C)$ [(or in the case of $T_r < T_C$, $\theta = -(T - T_C)/(T_r - T_C)$]. This procedure has been successfully tested in several metallic glasses as well as in polycrystalline magnetic solids [28,31]. However, if a secondary magnetic phase is present in the material, the temperature axis is rescaled in the form [30]:

$$\theta = -(T - T_C) / (T_{r1} - T_C) \qquad \text{for } T < T_C$$

$$\theta = (T - T_C) / (T_{r2} - T_C) \qquad \text{for } T > T_C \qquad (3),$$

where T_{r1} and T_{r2} are the temperatures, below and above T_C respectively, corresponding to the two reference points for each curve that correspond to $a \times \Delta S_M^{Peak}$ (with a in between 0 and 1). In this work a was selected to be 0.5. For this value, the reference points correspond to the half maximum of the $\Delta S_M(T)$ curves. The phenomenological $\Delta S_M(T)\backslash\Delta S_M^{Peak}$ curves were obtained using two reference temperatures. These curves for different magnetic fields collapse into the same curve for each sample in the paramagnetic zone (see figure 4 for the GdCeFe alloy). However, there are slight differences when comparing these master curves for all the alloys at the maximum magnetic field change (see the inset in figure 4).

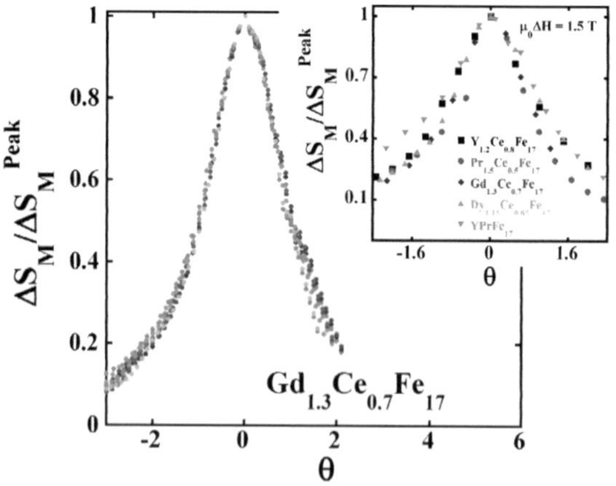

Figure 4. Master curve for $Gd_{1.3}Ce_{0.7}Fe_{17}$ obtained using two temperatures of reference (see text for details). The inset shows a comparison of the master curves for the studied pseudo-binary alloys under a magnetic field change $\mu_0\Delta H = 1.5$ T.

Conclusions

Room temperature x-ray and neutron powder diffraction confirm that the crystal structure of the $Y_{1.2}Ce_{0.8}Fe_{17}$, $Pr_{1.5}Ce_{0.5}Fe_{17}$, $Gd_{1.3}Ce_{0.7}Fe_{17}$, $Dy_{1.15}Ce_{0.85}Fe_{17}$ and $YPrFe_{17}$ pseudo-binary intermetallic compounds is rhombohedral with a space group $R\bar{3}m$. The values of the Curie temperature can be tuned between 250 and 295 K by small changes in the composition. In this way, it is possible to select the temperature of the maximum of $|\Delta S_M(T)|$. The Pr-containing compounds exhibit the higher maximum magnetic entropy change, as they have the largest saturation magnetization due to the collinear parallel configuration of both Fe and Pr sublattice magnetic moments. In addition, the calculated values for the *RC* can reach 65% of that for pure Gd.

Acknowledgments

Financial support from Spanish MICINN (research project MAT2011-27573-C04), the Basque Government (IT-347-07), Mexican CONACYT (CB-2010-01-156932), and Slovak R&D Agency (contract No. VVCE-0058-07) is acknowledged. We thank ILL and CRG D1B for allocating neutron beamtime. The liquid N_2 for the magnetic measurements was sponsored by the U.S. Steel Kosice.

References

[1] Buschow K H J 1977 *Rep. Prog. Phys.* **40** 1179.
[2] Brouha M *et al.* 1974 *IEEE Trans. Mag.* **10** 182.
[3] Piqué C *et al.* 2007 *Phys. Rev. B* **75** 224424.
[4] Mandal K *et al.* 2004 *J. Phys. D: Appl. Phys.* **37** 2628.
[5] Gorria P *et al.* 2008 *J. Phys. D: Appl. Phys.* **41** 192003.
[6] Álvarez P *et al.* 2010 *J. Phys.: Condens. Matter* **22** 216005.
[7] Álvarez P *et al.* 2012 *Phys. Rev. B* **86** 184411.
[8] *International Tables for Crystallography* 2005, ed T Hahn (IUCr, Springer, Dordrecht, The Netherlands).
[9] Gorria P *et al.* 2009 *Acta Mater.* **57** 1724.
[10] Tereshina I *et al.* 2005 *J. Alloys Compd.* **404** 172.
[11] Xiao Y G *et al.* 2006 *J. Alloys Compd.* **407** 1.
[12] Xiao Y G *et al.* 2006 *J. Alloys Compd.* **419** 15.
[13] Ben Kraiem M S *et al.* 2003 *J. Magn. Magn. Mater.* **256** 262.
[14] Richomme M S *et al.* 2010 *J. Alloys Compd.* **494** 5.
[15] Rodríguez Carvajal J 1993 *Physica B* **192** 55.
[16] Martinez-Blanco D *et al.* 2008 *J. Phys: Condens. Matter* **20** 335213.
[17] Arnold Z *et al.* 1994 *IEEE Trans. Mag.* **30** 619.
[18] Fujii H *et al.* 1995 *J. Alloys Compd.* **219** 10.
[19] Shen B G *et al.* 1998 *J. Phys. D: Appl. Phys.* **31** 2438.
[20] Wang J L *et al.* 2007 *Phys. Rev. B* **75** 174423.
[21] Cullity B D 1972, *Introduction to Magnetic Materials* (Addison-Wesley, Reading, MA).
[22] Isnard O *et al.* 1997 *J. Alloys Compd.* **262** 198.
[23] Grandjean F *et al.* 1999 *Solid State Comm.* **109** 779.
[24] Álvarez P *et al.* 2011 *Mat. Chem. Phys.* **131** 18.
[25] Álvarez Alonso P 2011, *Magnetocaloric and magnetovolume effects in Fe-based alloys*, PhD Thesis, (University of Oviedo, Oviedo , Spain).
[26] Franco V *et al.* 2006 *Appl. Phys. Lett.* **89** 222512.
[27] Álvarez P *et al.* 2011 *Intermetallics* **19** 982.
[28] Franco V *et al.* 2007 *Europhys. Lett.* **79** 47009.
[29] Franco V and Conde A 2010 *Int. J. Refrig.* **33** 465
[30] Franco V *et al.* 2009 *J. Magn. Magn. Mater.* **321** 1115.
[31] Álvarez P *et al.* 2010 *J. Alloys Compd.* **504S** S150.

Journal of Physics: Conference Series **549** (2014) 012020 doi:10.1088/1742-6596/549/1/012020

Successive magnetic transitions in $TbNiAl_2$ studied by neutron diffraction

D. P. Rojas[1], J. Sánchez Marcos[2] J. Rodríguez Fernández[3]

[1]Departamento de Física e Instalaciones-ETSAM, Universidad Politécnica de Madrid, Av. Juan Herrera 4, 28040, Madrid, Spain.

[2]Departamento de Química Física Aplicada. Facultad de Ciencias. Universidad Autónoma de Madrid. c/ Francisco Tomas y Valiente 7, 28049, Madrid, Spain.

[3]DCITIMAC, Facultad de Ciencias, Universidad de Cantabria, 39005, Santander, Spain.

E-mail: d.rojas@upm.es

Abstract. We report measurements of DC (AC) magnetic susceptibility and neutron diffraction on $TbNiAl_2$ alloy. The Rietveld refinements of the x-ray and neutron diffraction data are consistent with an orthorhombic structure of the type $MgCuAl_2$ (space group *Cmcm*). The results of DC (AC)- magnetic susceptibility show two successive magnetic transitions at 20 K and 11.7 K with antiferromagnetic and ferromagnetic (or ferrimagnetic) features, respectively. On the other hand, neutron diffraction patterns show that, below 20 K and down to 12 K, new reflexions appear, confirming the antiferromagnetic character of the transition observed in the macroscopic measurements. Also, at least one of these new reflexions, located at Q = 1.2 $Å^{-1}$, shifts to higher angles when the temperature decreases, indicating an incommensurate magnetic structure. Below 11 K, many reflexions disappear and new reflexions increase, evidencing a new magnetic transition.

1. Introduction

Tb intermetallic compounds have been extensively studied from both fundamental and applied points of view because their exotic magnetic behaviours and the interesting properties they show, associated for instance, with the magnetocaloric effect [1]. In particular, Tb alloys within the Tb-Ni-Al phase diagram present several magnetic transitions, which have been exhaustively studied. For instance, in TbNiAl two Tb sites out of three per unit cell order with a Neel temperature of T_N =47 K, while the third site is frustrated and contributes to a second magnetic phase transition at 23 K [2]. The magnetic structure of this alloy was determined by neutron diffraction [3]. Also, the maximum entropy change and refrigerant capacity obtained by means of thermal and magnetic properties are 13.8 $Jkg^{-1}K^{-1}$ near T_N and 494 Jkg^{-1} respectively for a magnetic field change of 50 kOe [4], values comparable to that observed in Gd.

On the other hand, in $TbNiAl_4$ antiferromagnetic (AFM) transitions, in zero applied magnetic field, were observed at 28 and 34 K. The magnetic structure of the lowest temperature phase is AFM with a *(0 1 0)* propagation vector, as obtained from neutron diffraction [5]. The magnetocaloric properties of single crystals also revealed interesting features with a large inverse entropy change associated with a field-induced step-like magnetic behaviour, and probably due to a transition from a commensurate AFM phase to incommensurate magnetic phase [6].

$TbNiAl_2$ is another ternary alloy of the Tb-Ni-Al diagram which is known to crystallize in orthorhombic $MgCuAl_2$- type of structure, and with a paramagnetic behaviour above 80 K [7]. However, details of the low temperature properties have not been reported yet. In the present work, we will provide a more detailed description of the magnetic properties of this alloy by DC and AC-susceptibility measurements, and by neutron diffraction.

2. Experimental details

A polycrystalline $TbNiAl_2$ alloy was prepared in an arc furnace from stoichiometric amounts of Tb (3N Alfa), Ni (5N, Alfa) and Al (5N, Alfa) metals. X-ray diffraction measurements were carried out in a Brucker diffractometer with CuK_α radiation. AC (DC) magnetic susceptibility measurements were collected in a Quantum Design PPMS in the temperature range 2 - 300 K and magnetic fields up to 9 T. Neutron diffraction measurements were recorded at the D1B and D1A beamlines, at the high flux reactor of the Institute Laue Langevin (ILL)-Grenoble.

3. Results and Discussion

Rietveld refinements of X-ray powder diffraction experimental data are consistent with an orthorhombic $MgCuAl_2$-type structure, space group *Cmcm*, and lattice parameters $a = 4.042(1)$ Å, $b = 10.263(1)$ Å and $c = 6.893(2)$ Å, in good agreement with those reported in the literature [7].

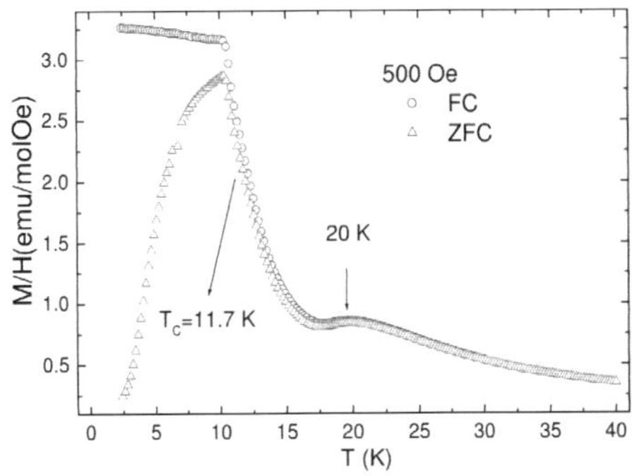

Figure 1. Temperature dependence of ZFC-FC curves of DC-magnetic susceptibility at 500 Oe.

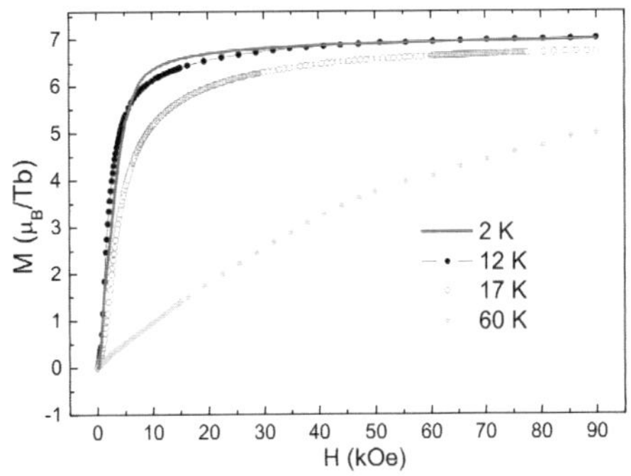

Figure 2. Field dependence of the isothermal magnetization curves at 2, 12, 17 and 60 K for $TbNiAl_2$ alloy.

The temperature dependence of zero field cooled (ZFC) and field cooled (FC) DC-magnetic susceptibility (χ=M/H) curves at 500 Oe is shown in Figure 1. At 20 K, the ZFC and FC curves present a small peak, and come together into one single curve, suggesting an AFM behaviour. At lower temperatures, below 12 K, the FC curve saturates on going to lower temperatures, which resembles a ferromagnetic (FM) behaviour. From the inflexion point of these curves, an ordering temperature of 11.7 K is estimated. The presence of a maximum in the ZFC curve (around 10 K), and a large irreversibility, point towards the existence of FM-like domains in this material. In the high temperature region (above 50 K), a Curie-Weiss behaviour is found, with $\mu_{eff} = 9.37$ μ_B, value near to the free ion Tb^{3+} (9.72 μ_B), and a negative paramagnetic Curie temperature (θ_p=-6 K) suggesting the presence of AFM interactions. Also, the field dependence of the isothermal magnetization at 2, 12 K and 17 K (see Figure 2), shows a full saturation at the highest achieved magnetic field of 90 kOe, with values of 7.02 μ_B (2 and 12 K) and 6.75 μ_B (17 K) still far from the expected gJ value of 9 μ_B for the full multiplet of Tb^{3+} (g=3/2, J=6), which could be attributed to crystalline field effects.

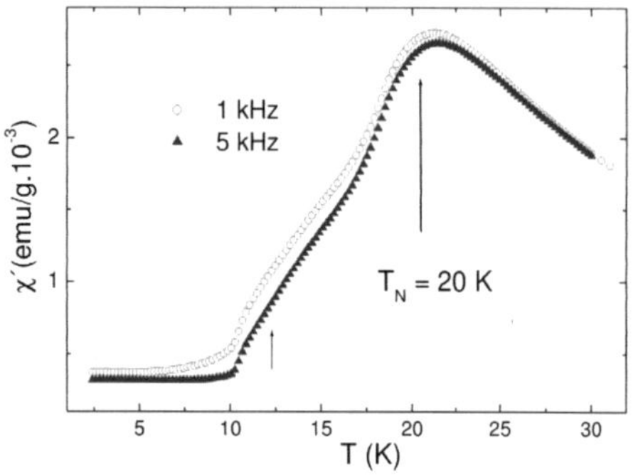

Figure 3. Temperature dependence of the real component (χ') of the AC-magnetic susceptibility at 1 and 5 kHz.

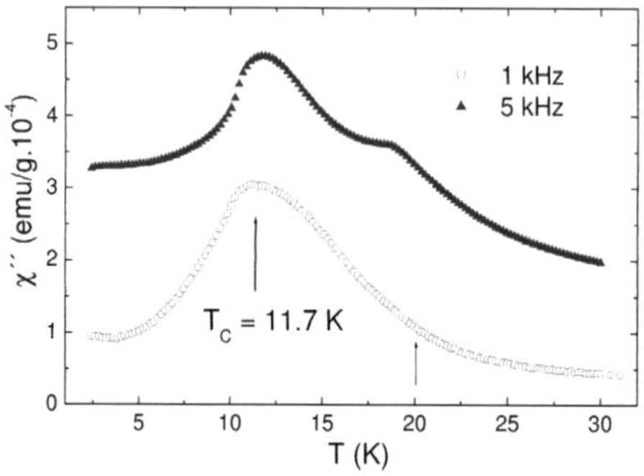

Figure 4. Imaginary component (χ'') vs T of the AC-magnetic susceptibility at 1 and 5 kHz.

Measurements of the AC-susceptibility also could provide an useful information about the magnetic transitions. In this sense, the temperature dependence of the real (χ') and imaginary (χ'') components of the AC-susceptibility at frequencies of 1 and 5 kHz are presented in figure 3 and 4, respectively. In χ', a maximum at 20 K, and a broad anomaly centered around 12 K are observed. Moreover, in (χ''), a peak around 11.7 K appears, whereas at 20 K a second *tiny* anomaly develops for 5 kHz. As it is well known the peak in χ'' is commonly associated to dynamic processes of domain movements, whereas the increase of the absolute value of χ'' for 5 kHz can be attributed to a contribution to the susceptibility arising from currents induced in the sample by the alternating magnetic field, as observed in other intermetallic systems [8]. Therefore, the overall behaviour of the AC-magnetic susceptibility is in good agreement with the DC-magnetic susceptibility results indicating an AFM transition at 20 K and a FM (or ferrimagnetic) ordering around 11.7 K.

In figure 5, neutron diffraction patterns at different temperatures of the $TbNiAl_2$ alloy are presented. A comparison of the pattern in the paramagnetic region (30 K) with those related to the low temperature magnetic phases at 2 and 14 K, are shown in figure 5(a,b). It is observed that new peaks appear in the low temperature diffraction patterns, indicating an AFM arrangement of the magnetic moments. We could notice that at 14 K new peaks are detected up to 65° whereas at 2 K a new set of magnetic reflexions appear even up to 130°. In both magnetic phases it is observed a reduction of the background due to the decrease of the anisotropic scattering of the magnetic moments with the arising of the ordering state. An interesting feature is found at temperatures between 17.5 and 11 K, where the magnetic peak at $Q = 1.2$ Å^{-1} shifts to higher angles when the temperature decreases, indicating an incommensurate magnetic phase (figure 5 (c)). Due to the complexity of the magnetic structures, it has not been possible to determine the appropriate magnetic propagation vectors. Although, the overall evolution of the magnetic Bragg peaks with the temperature, is consistent with the onset of the magnetic transitions, as estimated from the DC and AC-magnetic susceptibilities (figure 5 (d)).

From the above results it is evident the existence of two magnetic transitions of different nature in the $TbNiAl_2$ alloy. This situation is not uncommon as it is similar to that found in other ternary alloys as TbNiAl [2, 3, 4] and $TbNiAl_4$ [5, 6].

4. Conclusions

Summarizing, the results of DC(AC)-susceptibility and neutron diffraction on the $TbNiAl_2$ alloy are consistent with two successive magnetic transitions at T_N = 20 K and T_C= 11.7 K. The

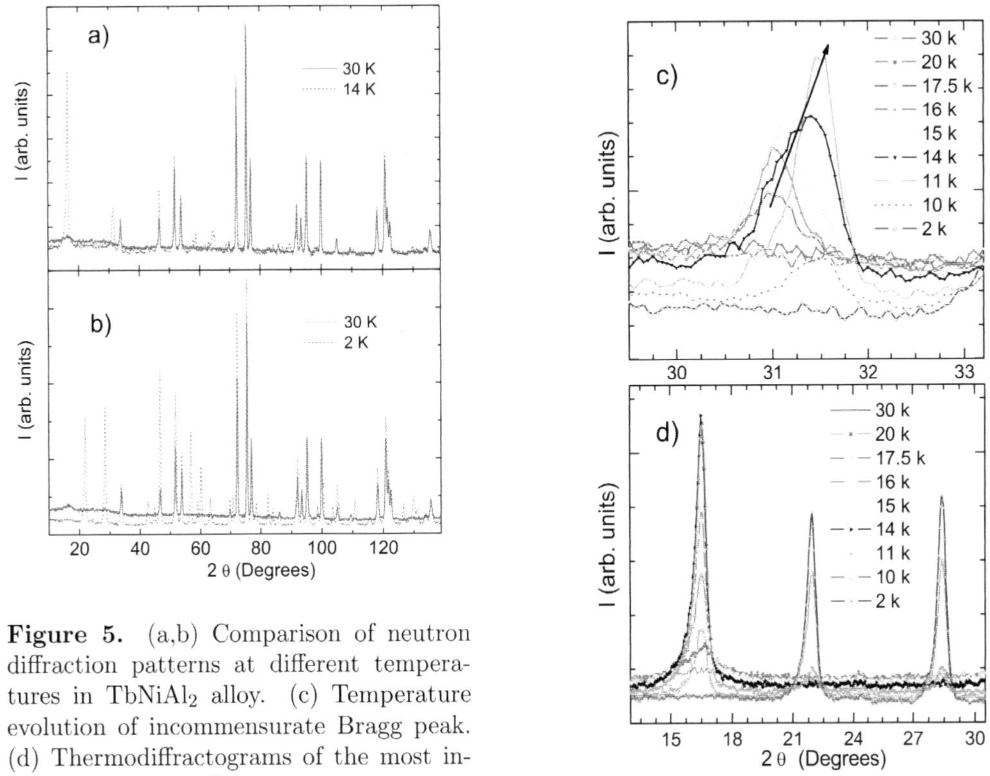

Figure 5. (a,b) Comparison of neutron diffraction patterns at different temperatures in $TbNiAl_2$ alloy. (c) Temperature evolution of incommensurate Bragg peak. (d) Thermodiffractograms of the most intense magnetic peaks.

analysis of thermodiffractograms of neutron diffraction patterns indicates an incommensurate AFM structure for the first, and very likely, a commensurate ferrimagnetic ones for the second at lower temperatures, respectively. It would be interesting to study the influence of pressure and magnetic field on the magnetic transitions of this intermetallic, as well as the magnetocaloric effect.

Acknowledgments

This work was supported by MAT 2008-06542-C04 and MAT2011-27573-C04 projects. The authors thank ILL and CRG-D1B for allocating neutron beam time.

References

[1] Pecharsky V K and Gschneidner K A Jr 2007 *Handbook of Magnetism and Advanced Magnetic Materials* vol 4, ed H Kronmüler and S Parkin (Chichester: Wiley).
[2] Javorsky P, Burlet P, Sechovsky V, Andreev A V, Brown J, and Svoboda P 1997 *J. Magn. Magn. Mater.* **166** 133
[3] Maletta H and Sechovsky V 1994 *J. Alloys Compd.* **207/208** 254
[4] Niraj K Singh, Suresh K G, Nirmala R, Nigam A K and Malik S K 2006 *J. Magn. Magn. Mater* **302** 305.
[5] Hutchison W D, Goossens D J, Nishimura K, Mori K, Isikawa Y and Studer A J 2006 *J. Magn. Magn. Mater* **301** 352.
[6] Lingwei Li, Katsuhiko Nishimura and Hutchison W D 2009 *Solid State Commun.* **149** 932.
[7] Romaka V A, Zarechnyuk O S, Rykhal R M, Yarmolyuk Y P and Skolozdra R V 1982 *Fizika Metallov i Metallovedenie* **54** (2) 410.
[8] Rojas D P, Rodríguez Fernández J, Espeso JI and Gómez Sal J C 2009 *Physica B* **404** 2938.

Journal of Physics: Conference Series **549** (2014) 012021 doi:10.1088/1742-6596/549/1/012021

Determination of the crystal and magnetic structure of the $DyCrO_4$ -scheelite polymorph by neutron diffraction

A. J. Dos santos-García[1], E. Climent-Pascual[1], M. G. Rabie[1], J. Romero de Paz[1], J. M. Gallardo Amores[1], D. Khalyavin[2] and R. Sáez-Puche[1]

[1]Departamento de Química Inorgánica, Facultad de Químicas, Universidad Complutense de Madrid, 28040 Madrid, Spain.

[2]ISIS Facility, Rutherford Appleton LaboratoryCCLRC, Chilton, Didcot, Oxfordshire OX11 0QX, United Kingdom

E-mail: `rsp92@ucm.es`

Abstract. Neutron diffraction data of $DyCrO_4$ oxide, prepared at 4 GPa and 833 K from the ambient pressure zircon-type, reveal that crystallize with the scheelite-type structure, space group $I4_1/a$. Accompanying this structural phase transition induced by pressure the magnetic properties change dramatically from ferromagnetism in the case of zircon to antiferromagnetism for the scheelite polymorph with a $T_N = 19$ K. The analysis of the neutron diffraction data obtained at 1.2 K has been used to determine the magnetic structure of this $DyCrO_4$-scheelite oxide which can be described with a k = [0, 0, 0] as propagation vector, where the Dy and Cr moments are lying in the *ab*-plane of the scheelite structure. The ordered magnetic moments are 10 μ_B and 1μ_B for Dy^{+3} and Cr^{+5} respectively.

1. Introduction

Phase transitions induced by pressure have been studied for the zircon mineral $ZrSiO_4$ and the analogous zircon-types of RXO_4 (X= As and V) which transform the tetragonal space group $I4_1$/amd into a scheelite-type structure, space group $I4_1/a$. Recently for $YbPO_4$ and $LuPO_4$, it was revealed using in-situ synchrotron x-ray diffraction experiments that the pressure-induced zircon-to-scheelite transition is reversible and the scheelite high-pressure phase cannot be quenched to ambient conditions in these phosphates [1]. In the case of vanadates and arsenates, this phase transition from zircon to scheelite, with an increase in the density of 11%, takes place at 8 GPa and 773 K [2, 3]. Regarding the high pressure phase transformations in $RCrO_4$, in which we are interested, only few studies have been reported. This could be due to the high instability of Cr^{+5} and its tendency to be reduced to give the most stable distorted phase of perovskite $RCrO_3$. However, it has been recently reported by our research group that zircon-$RCrO_4$ phases transform to a scheelite polymorph at 4 GPa and 813 K for the $HoCrO_4$ and $TbCrO_4$ compounds, which are quenchable after releasing the pressure [4, 5]. Accompanying this structural phase transition from zircon to scheelite the magnetic properties change dramatically. In this sense, most of the zircon-type $RCrO_4$ oxides being ferromagnetic become antiferromagnetic for the corresponding known scheelite polymorph. Detailed structural analysis by x-ray diffraction

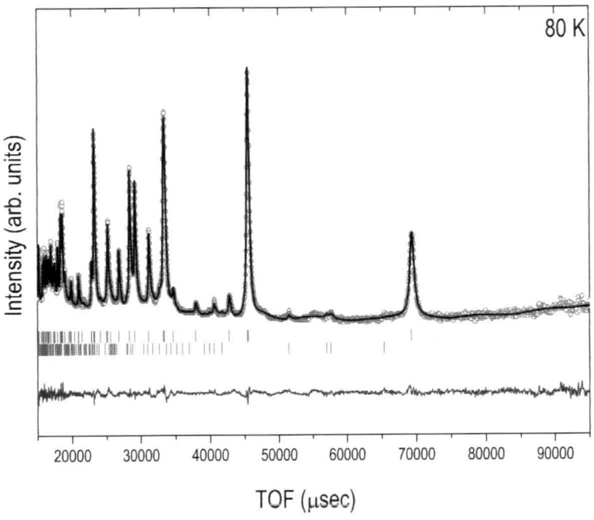

Figure 1. Rietveld refinement of the $DyCrO_4$ scheelite polymorph at 80 K for TOF neutron diffraction data. First raw of the vertical bars (i) corresponds to the $DyCrO_4$ scheelite. Second line (ii) is for $DyCrO_3$ distorted perovskite.

(XRD) and neutron diffraction (NPD) allow us to establish structure-magnetic properties relationships, which represents one of the main interests in the solid-state science. In this work, we aim to study the scheelite-$DyCrO_4$ oxide through investigating the structural and magnetic properties. Therefore, we focus first on the structural phase transition from the zircon-like to the scheelite-like structure under pressure and temperature. Second, we report the study of the magnetic properties of this $DyCrO_4$ from magnetization and neutron diffraction measurements.

2. Experimental

The $DyCrO_4$ zircon-phase sample, which was prepared by a precursor method described elsewhere, has been treated at 4 GPa and 833 K and 35-50 minutes in a CONAC-type press. $DyCrO_4$ scheelite oxide was first studied by X-ray powder diffraction using a Philips X'Celerator diffractometer with a primary beam monochromator consisting of a curved Ge (1 1 1) crystal aligned to obtain $CuK_{\alpha 1}$ radiation. A range of 2θ-angle between 10° and 90° was recorded with a step size of 0.0167° collected for 150s. Neutron diffraction experiments were done at the ISIS facility in the UK using the WISH cold-neutron time of flight (TOF) powder diffractometer. All obtained diffraction patterns were analyzed using the FullProf Suite Program. Magnetic susceptibility and magnetization measurements were performed in a Squid Quantum Design MPS XL from 300 K to 2 K temperature range in different magnetic fields up to 5T.

3. Results and discussion

3.1. Structural characterization

The structure of the $DyCrO_4$ compound obtained after treating the $DyCrO_4$-zircon polymorph at 4 GPa and 833 K has been refined from the X-ray and neutron diffraction data using the Rietveld method with the Fullprof program. As it is shown in Figure 1 the neutron diffraction data obtained at 80 K show the reflections conditions compatible with the space group $I4_1/a$ characteristic of the scheelite type structure.

The refined atomic positions are Dy at 4b (0, 1/4, 5/8), 4a (0, 1/4, 1/8) for Cr and the oxygen atoms are located at the 16f (x, y , z). The lattice parameters and main bond angles and interatomic distances are given in Table 1.

It is worth noting that small amount of impurities such as $DyCrO_3$ has also been include in the refinement, 8.98 (57) %. The zircon-type structure of $RCrO_4$ can be described as formed by

Table 1. Control sequences to describe lines and symbols in figure captions.

T (K)	80	1.2
a (Å)	5.0170(2)	5.0148(2)
b (Å)	11.3079(6)	11.3037(5)
V (Å^2)	284.62(2)	284.27(2)
U_{Dy} (Å^2)	0.0003(8)	0.0002
U_{Cr} (Å^2)	0.019(3)	0.0002
x_o	0.2437(5)	0.2437
y_o	0.6091(4)	0.6091
z_o	0.5486(2)	0.5486
U_o (Å^2)	0.0105(14)	0.0002(8)
d_{Cr-O} (Å)	1.7020(22) x 4	1.7020 x 4
d_{Dy-O} (Å)	2.3413(22) x 4	2.3413 x 4
	2.4177(23) x 4	2.4177 x 4
Cr - O - Dy (deg.)	121.759 (114)	121.759
	132.242(82)	132.242
$\mu_{xy,Dy}$ (μ_B)	——	9.67(10)
$\mu_{xy,Cr}$ (μ_B)	——	0.99(12)
R_{Bragg} (%)	2.44	5.49
R_p (%)	3.81	4.02
R_{wp} (%)	2.10	5.18
χ^2	2.62	1.32
Magnetic R-factor (%)	——	6.82

chains of CrO_4 and RO_8 polyhedra which are parallel to the c-axis and alternating by an edge sharing as seen in Figure 2a. Adjacent rods are connected along a and b-axis by zig-zag of edge sharing between RO_8 polyhedra. In case of the scheelite structure, CrO_4 distorted tetrahedra are aligned along the a-axis, while RO_8 bisdisphenoids are intercalated between the CrO_4 tetrahedra (Figure 2b).

The RO_8 polyhedra shares edges with adjacent RO_8 polyhedra and form along a-direction zig-zag chains, which are cross-linked by tetrahedra via corner-sharing (figure 2b). However, there is no edge-sharing between RO_8 and CrO_4 polyhedra. The difference in the tilting angle between different polyhedra, as well as in the packing after the transition provides an remarkable change in the Cr- O - R angles as it will be discussed later to explain the different magnetic behavior that presents this scheelite polymorph in comparison with the zircon one.

3.2. Magnetic properties

Figure 3 shows the evolution of the magnetic susceptibility with the temperature and it can observed that the susceptibility obeys a Curie- Weiss behavior, $\chi = C/T + \theta$ in the temperature range 300 - 40 K, see Figure 3 inset.

It is also observed a small anomaly at 150 K due to the presence of the distorted perovskite $DyCrO_3$ that become antiferromagnetically ordered at this temperature with a small canting. However the presence of this phase doesn't affect to the magnetic properties of the scheelite-$DyCrO_4$ oxide. In this sense, the obtained magnetic moment takes the value of 11.2 μ_B which fairly agrees with the expected for this compound is 10.80 μ_B. The Weiss constant (θ) takes

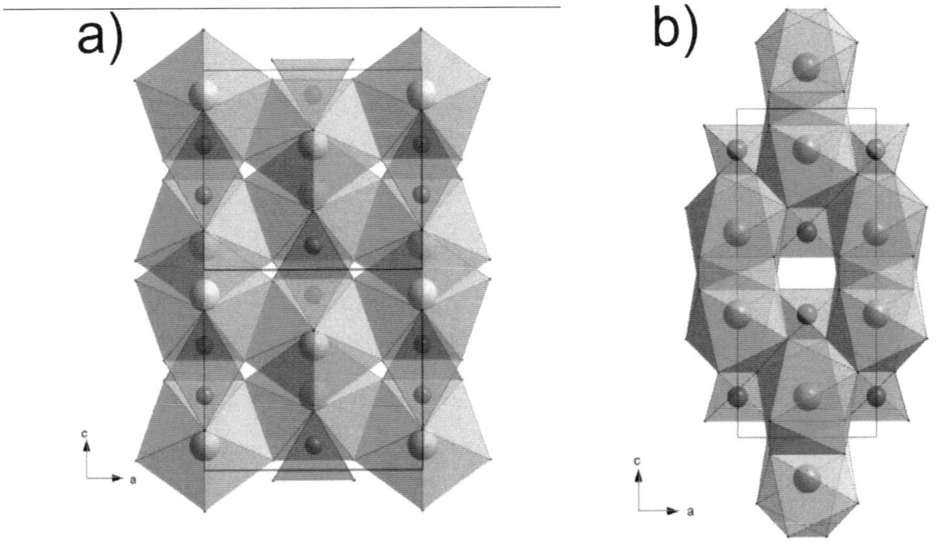

Figure 2. a) Crystal structure of the zircon-$DyCrO_4$ and b) scheelite-$DyCrO_4$ polymorph.

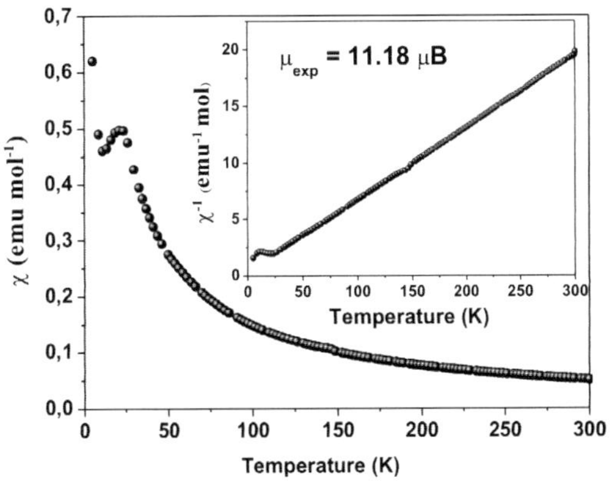

Figure 3. ZFC magnetization measurements for $DyCrO_4$ scheelite in a 0.05 T field. The inset shows the temperature dependence of the reciprocal susceptibility.

the value of -5.6 K which is indicative of the existence of antiferromagnetic interactions in this oxide. These interactions are fully confirmed the onset of a net maximum at 20 K in the χ vs T plot, see Figure 3. The increasing of the susceptibility below the T_N is due to presence of the small amount of impurities mentioned earlier. The different magnetic behavior that present both $DyCrO_4$ polymorphs, the scheelite form is antiferromagnetic, while the zircon-$DyCrO_4$ have been reported to be ferromagnetic [6] can be explained from the analysis of the structure of both polymorphs. As it was reported previously, these interactions take place through the superexchange Cr^{+5}-O-Dy^{+3} pathway in which both Cr and Dy are involved. Since the Cr-O and Dy-O interatomic distances remain almost constant for both polymorph, the change in the sign of the exchange integral going from the zircon to scheelite form is mainly due to remarkable differences found in the Cr^{+5}-O-Dy^{+3} bond angle found in both polymorphic phases. The magnetic structure of the $DyCrO_4$-scheelite polymorph has been determined from the analysis of the neutron diffraction data obtained at 1.2 K. The neutron powder diffraction patterns

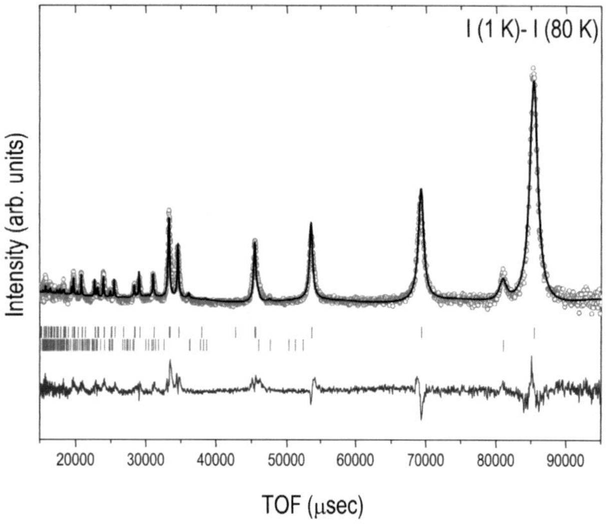

Figure 4. Magnetic structure fit from the difference between the 1.2 and 80 K neutron diffraction profiles of $DyCrO_4$. First ticks of vertical marks (i) corresponds to the magnetic contribution from the $DyCrO_4$ scheelite and second line (ii), magnetic contribution from the $DyCrO_3$ distorted perovskite.

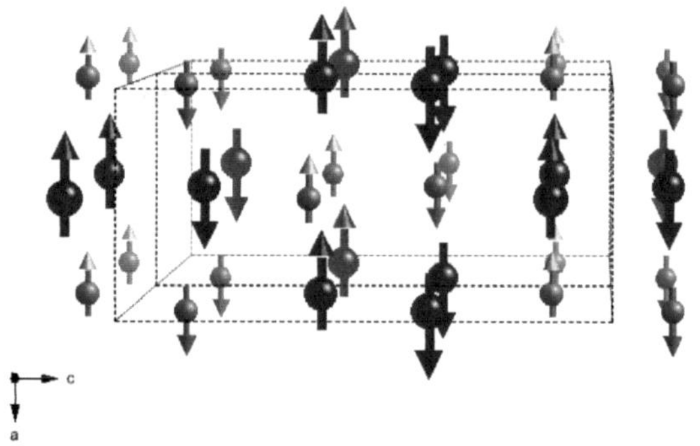

Figure 5. Magnetic structure of the $DyCrO_4$ scheelite. Blue and yellow arrows stem for the Dy and Cr magnetic moments respectively.

show the onset of new reflections forbidden for the space group $I4_1/a$ along with a progressive increase of certain reflections below $T_N = 20$ K. These magnetic reflections, see Figure 4, have been indexed on the basis of the coincidence between the crystal and magnetic cells, with the propagation vector k = [0, 0, 0].

The thermal evolution of the integrated intensity of the (002) and (110) magnetic reflections reveals that these magnetic reflections disappear at 22 K, a higher ordering temperature than the observed one from the magnetic susceptibility measurements. The magnetic structure is defined in the primitive P1 space group and the best agreement (RM= 0.068) between the experimental and calculated neutron diffraction profiles was obtained when Cr^{+5} and Dy^{+3} magnetic moments are aligned parallel to the ab-plane in the same manner like for $GdCrO_4$ scheelite [7],see Figure 5.

The saturation magnetic moments obtained from the Rietveld refinement are $9.67(10)\mu_B$ and $0.99(12)\mu_B$ for Dy^{+3} and Cr^{+5} respectively. As it can be observed in Figure 5 the magnetic moments of Dy^{+3} are antiferromagnetically coupled between the (Dy_2O_{14}) dimers resulting from the two (DyO_8) bisdiphenoids sharing a edge in the ab-plane of the scheelite structure. The Cr^{+5} moments are also antiferomagnetically coupled in the ab-plane of the structure given as result the collinear magnetic structure depicted in Figure 5.

4. Acknowledgments

The authors acknowledge funding from Comunidad de Madrid under the research project S-2009/PPQ-1626 and MICINN through MAT2010-19460. AJD is grateful to MICINN and UCM for a Juan de la Cierva contract (JCI-2012-08229). M. Rabie is also grateful to financial support for a pre-doctoral grant under research SOPRANO-Marie Curie actions (FP7/20072013 under Grant Agreement no.214040). Authors are also indebted to ISIS-facility for beamtime allocation at WISH instrument and Dr. L. C. Chapon and Dr. P. Manuel for valuable comments.

5. References

[1] Zhang F, Lang M, Ewing R, Lian J, Wang Z, Hu J and Boatner L 2008 *Journal of Solid State Chemistry* **181** 2633 – 2638

[2] Duclos S J, Jayaraman A, Espinosa G, Cooper A and Sr R M 1989 *Journal of Physics and Chemistry of Solids* **50** 769 – 775

[3] Stubican V S and Roy R 1963 *Journal of Applied Physics* **34** 1888–1890

[4] Climent-Pascual E, de Paz J R, Gallardo-Amores J M and Sez-Puche R 2007 *Solid State Sciences* **9** 574 – 579

[5] Climent E, Gallardo J, de Paz J R, Taira N and Puche R S 2009 *Journal of Alloys and Compounds* **488** 524 – 527

[6] Long Y, Liu Q, Lv Y, Yu R and Jin C 2011 *Phys. Rev. B* **83**(2) 024416

[7] Dos Santos-García A J, Climent-Pascual E, Gallardo-Amores J M, Rabie M G, Doi Y, de Paz J R, Beuneu B and Saez-Puche R 2012 *Journal of Solid State Chemistry* **194** 119 – 126

Neutron structural characterization and transport properties of the oxidized and reduced $LaCo_{0.5}Ti_{0.5}O_3$ perovskite oxide.

R. Martinez-Coronado[1], A. Aguadero[2], J.A. Alonso[1], D. Perez-Coll[3], M.T. Fernandez-Díaz[4].

[1]Instituto de Ciencia de Materiales de Madrid, C.S.I.C., Cantoblanco E-28049 Madrid.
[2]CIC Energigune, Albert Einstein 48, 01510 Alava, Spain
[3]Instituto de Cerámica y Vidrio-CSIC, Cantoblanco, 28049 Madrid, Spain
[4]Institut Laue Langevin, BP 156X, Grenoble, F-38042, France

E-mail: rmartinez@icmm.csic.es

Abstract. Polycrystalline oxygen-stoichiometric $LaCo_{0.5}Ti_{0.5}O_3$ perovskite oxide has been prepared by soft-chemistry procedures followed by annealing in air at 800ºC. A new reduced $LaCo_{0.5}Ti_{0.5}O_{3-\delta}$ specimen has been obtained by topotactical oxygen removal in an H_2/N_2 (5%/95%) flow at 600ºC. The structural characterization has been conducted from neutron powder diffraction (NPD) data, very sensitive to the contrast between Co and Ti and the oxygen stoichiometry. Both perovskites (oxidized and reduced) crystallize in the orthorhombic *Pbnm,* space group. The partial reduction of Ti^{4+} to Ti^{3+} in the reduced phase is accompanied with the occurrence of oxygen vacancies, located at the axial octahedral sites, and it is expected to support the ionic conductivity, as usually observed in oxygen-defective perovskites. Thermogravimetric analysis (TGA) substantiates the oxygen stoichiometry and the stability range of the reduced sample. All the samples in study display a semiconductor-like behavior with values that not reach below to 0.5 Scm^{-1} for all the phases. Moreover, the measured thermal expansion coefficients perfectly match with the values usually displayed by SOFC electrolytes.

1. Introduction

Materials with perovskite-like structure are receiving great attention due to their vast range of possible applications. Perovskite-type oxides are widely used as functional materials in ferroelectric, thermoelectric, dielectric, magnetoelectronic devices and in solid oxide fuel cells (SOFC), among wide panoply of applications. Their unique properties derive from an unmatched chemical flexibility in spite of their relatively simple crystal structure [1]. The reduction of stoichiometric phases into novel oxygen hypo-stoichiometric oxides is a powerful tool for the development of new materials with novel magnetic or transport properties [2]. Moreover, the stabilization of transition-metal perovskites with an adequate concentration of oxygen vacancies under reducing atmosphere can also be of application in new mixed electronic-ionic conductors for energy-conversion devices such as solid oxide fuel cells, oxygen separation membranes or solid oxide electrolyzers.

Owing to many interesting properties such as rich structural characteristics, transport and magnetic properties [3-6], perovskites oxides of R(III)Ti(III)O_3 (R= rare earth) have been one of the hot subjects of recent studies. Among perovskite oxides that exhibit appealing properties, ceramic titanates are recognized to present high stability under reducing environments with high tolerance to sulphur poisoning. The formation of R(M,Ti)O_3 (R= rare earth; M= metal transition) is may be feasible either for trivalent M and Ti ions (having in mind that Ti^{3+} ions are difficult to stabilize) or with a combination of divalent M and tetravalent Ti ionic states, under the appropriate atmosphere during the high-temperature synthesis [7]. When Co(III) ions are substituted by Ti(IV), as reported in the $LaCo_{1-x}Ti_xO_3$ solid solution [8], a mixed valence system is obtained as the charge neutrality requires the partial reduction to divalent Co ions.

$LaCo_{0.5}Ti_{0.5}O_3$ perovskite has been previously reported by several researchers, which defined the crystal structure by neutron powder diffraction in both orthorhombic and monoclinic unit-cell [9-11]; also x-ray diffraction, thermogravimetric analysis and magnetic susceptibilities were used to characterize this material [12]. Although there is a comprehensive study of this perovskite, the research of the reducibility has not been yet considered.

In this work we have analyzed the reducibility of the $LaCo_{0.5}Ti_{0.5}O_3$ perovskite and the crystal structure evolution from the oxidized to the reduced phase. The analysis of NPD data for both oxidized and reduced specimens suggests the presence of Ti^{3+}-Ti^{4+} mixed valence in the reduced phase. The characterization has been completed with thermal expansion, electric conductivity measurements and thermal analysis under oxidizing and reducing atmosphere.

2. Experimental

$LaCo_{0.5}Ti_{0.5}O_3$ perovskite was prepared as a polycrystalline powder from citrate precursors obtained by soft-chemistry procedures. Stoichiometric amounts La_2O_3 (pre-dried at 900ºC), $Co(NO_3)_2{\cdot}6H_2O$ and $TiC_{10}H_{14}O_5$ were solved in citric acid and some drops of nitric acid. The solution was then slowly evaporated, in order to favor the dissolution of the rare-earth oxide, leading to organic resins that contain a homogeneous distribution of the involved cations. The formed resins were dried at 120 ºC and decomposed at 600 ºC for 12 h, heating with a 50 ºC·h^{-1} ramp, in air. All the organic materials and nitrates were eliminated in a subsequent treatment at 800°C in air, for 2 hours, which gave rise to the pure perovskite oxide phase. The reduced $LaCo_{0.5}Ti_{0.5}O_{3-\delta}$ perovskites were prepared by treating the oxidized phase under an 5%H_2/95%N_2 flow (60 mL min^{-1}) at 600 ºC for 4 h in alumina boats.

The initial characterization of the product was carried out by XRD with a Bruker-axs D8 Advanced diffractometer (40 kV, 30 mA), controlled by a DIFFRACT^{PLUS} software, in Bragg-Brentano reflection geometry with Cu K_α radiation (λ = 1.5418 Å) and a PSD (Position Sensitive Detector). A filter of nickel allows the complete removal of Cu K_β radiation. For the structural refinement NPD patterns were collected at the D1A diffractometer of the ILL, Grenoble, with a wavelength λ= 1.910 Å at room temperature. About 2 g of the sample were contained in a vanadium can and placed in the isothermal zone of a furnace with a vanadium resistor operating under vacuum ($P_{O2} \approx 10^{-6}$ Torr), and the counting time was 2 h per pattern in the high-intensity mode. The NPD data were analyzed by the Rietveld method [13] with the FULLPROF program [14]. A pseudo-Voigt function was chosen to generate the line shape of the diffraction peaks. The following parameters were refined in the final run: scale factor, background coefficients, zero-point error, pseudo-Voigt corrected for asymmetry parameters, positional coordinates and isotropic thermal factors for all the atoms. The coherent scattering lengths for La, Co, Ti and O were 8.24, 2.49, -3.438 and 5.803 fm, respectively [14].

Thermal analysis was carried out in a Mettler TA3000 system equipped with a TC10 processor unit. Thermogravimetric (TG) curves were obtained in a TG50 unit, working at a heating rate of 10 °C min^{-1}, in a reducing H_2(5%)/N_2(95%) flow of 0.3 L min^{-1}. The heating rate was 10 °C min^{-1}, using about 50 mg of sample in each experiment.

Measurements of the thermal expansion coefficient and electrical conductivity required the use of sintered samples. The obtained density is around 90-95%. Thermal expansion of the sintered samples was performed in a dilatometer Linseis L75HX1000, between 300 and 800 ºC in air and

H_2(5%)/N_2(95%) atmospheres. The conductivity was measured between 25 and 850 °C in the requested atmosphere, by the four-point method in bar-shaped pellets under DC currents between 0.05 and 0.10 A. The currents were applied and collected with a Potenciostat-Galvanostat AUTOLAB PGSTAT 302 from ECO CHEMIE.

3. Results and Discussion

3.1. Crystal structure

The oxidized and reduced $LaCo_{0.5}Ti_{0.5}O_3$ perovskites were obtained as well-crystallized powders. Single-orthorhombic perovskite phases were identified from laboratory XRD (Fig. 1) for the reduced and oxidized sample. No impurity phases were detected.

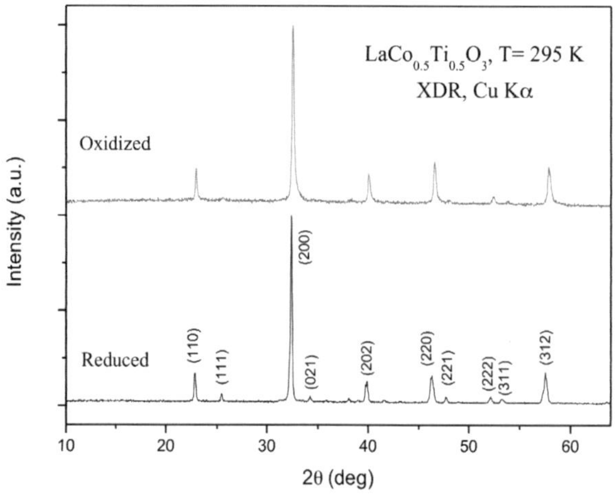

Fig. 1. (Color online) XRD patterns with Cu Kα radiation for $LaCo_{0.5}Ti_{0.5}O_3$, characteristic of pure orthorhombic perovskite phases.

To carry out a more accurate structural study of the $LaM_{0.5}Ti_{0.5}O_3$ (M= Co and Ni) oxides, we performed a NPD investigation at room temperature for all the perovskites; neutrons are especially sensitive to the nature of these atoms since they show very contrasting (positive for Co and negative for Ti) scattering lengths. The crystal structure of the oxidized and reduced $LaCo_{0.5}Ti_{0.5}O_3$ is defined in the orthorhombic *Pbnm* space group (No. 62), Z=4, as was previously reported by Clairns et al. [11]. La and O1 atoms are located at 4*c* (x,y,1/4) positions, M and Ti distributed at random at 4*b* (1/2,0,0), and oxygen atoms O2 at 8*d* (x,y,z). Therefore, for both oxidized and reduced phases the Co and Ti atoms are randomly distributed and no crystallographic long-range order was observed.

The refinement of the occupancy factors of the oxygen atoms for the oxidized phase led to a full stoichiometry, while the reduced phase shows oxygen vacancies concomitant with the presence of Ti^{3+}. The vacancies are concentrated at O1 sites (axial oxygen atoms); O2 showed occupancies slightly higher than 1 and was then fixed to unity. The refined occupancy factors of oxygen atoms for the reduced phase lead to the $LaCo_{0.5}Ti_{0.5}O_{2.91(1)}$ stoichiometry. Fig. 2 illustrates the good agreement between the observed and calculated NPD patterns for the oxidized and reduced $LaCo_{0.5}Ti_{0.5}O_3$ at room temperature.

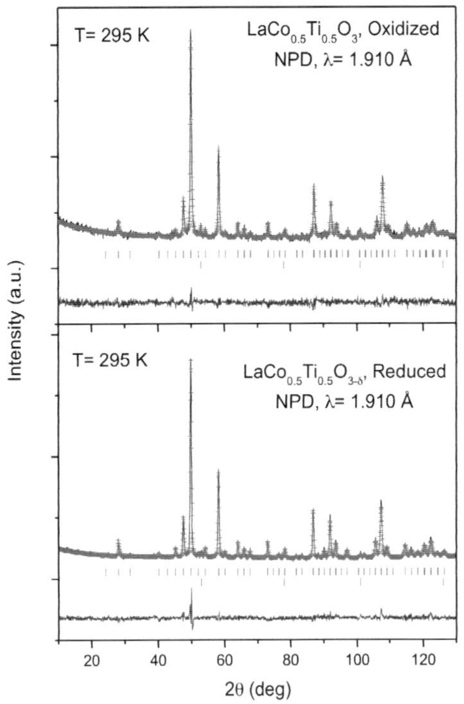

Fig. 2. (Color online) Observed (crosses), calculated (full line) and difference (at the bottom) NPD profiles for reduced and oxidized $LaCo_{0.5}Ti_{0.5}O_3$ at 295 K, refined in the orthorhombic *Pbnm* space group. The two series of tick marks correspond to the positions of the allowed Bragg reflections for the main phase and vanadium.

An alternative refinement of the crystal structure of the oxidized and reduced samples was carried out in the monoclinic $P2_1/n$ space group; in this model the Rietveld refinements display the existence of a certain level of anti-site disorder between Co and Ti cations. However, the results of the refinement at $P2_1/n$ are similar to those obtained with the *Pbnm*; the R_{Bragg} discrepancy factor is practically the same (10%) and the displacement B factors for Zn and Mn become unrealistically high (around 5 $Å^2$); therefore this model was discarded. This is in disagreement with the results by Rodriguez [10], who described an monoclinic $P2_1/n$ symmetry for the La(Co,Ti)O_3 from NPD data.
Previous reports of the $LaCo_{0.5}Ti_{0.5}O_3$ perovskite have achieved similar R_{Bragg} discrepancy factor for the disordered model in *Pnma* and the ordered model in $P2_1/n$ [10,11]. Refinements in the ordered model revealed small amounts of intermixing (5%) of Co and Ti on the B and B′ sites. The refinements show that the higher-temperature synthesis (1300 °C) method gives more complete B site ordering than was found for the lower-temperature synthesis is temperature (900 °C) method previously reported [10].

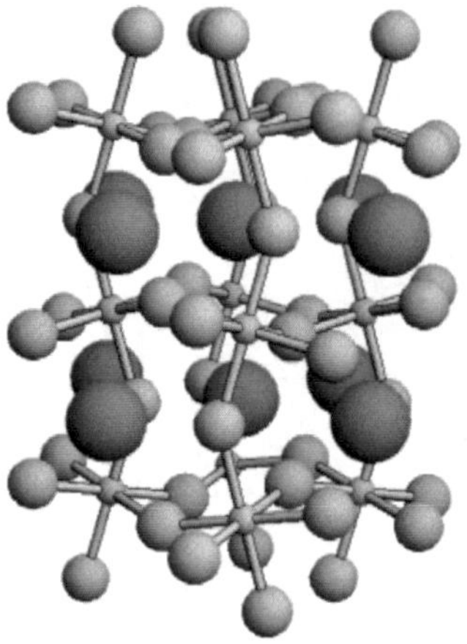

Fig. 3. Schematic view of the orthorhombic crystal structure of $LaCo_{0.5}Ti_{0.5}O_3$.

Fig. 3 sketch the crystal structure observed in both the oxidized and reduced phases, which is defined in the orthorhombic *Pbnm* ($a^-a^-c^+$) space group. Table I summarizes the unit-cell, atomic, thermal parameters and discrepancy factors after the Rietveld refinements of all the samples at room temperature. The cell volume of the reduced phase (242.92(1) $Å^3$) are expanded with respect to the oxidized sample (240.29(1) $Å^3$), as expected from the larger size of Ti^{3+} (i.r.: 0.650 Å) *vs* Ti^{4+} (i.r.: 0.605 Å) [15].

Table I. Unit-cell and thermal parameters for the oxidized and reduced $LaM_{0.5}Ti_{0.5}O_3$ (M= Co and Ni) in orthorhombic *Pbnm* (No. 62) space group, from NPD at 295K. La and O1 at 4*c* (x,y,1/4) positions, Co/Ti at 4*b* (1/2,0,0), and O2 at 8*d* (x,y,z).

M	Co (oxidized)	Co (reduced)
a (Å)	5.5457(2)	5.5605(1)
b (Å)	5.5344(2)	5.5608(2)
c (Å)	7.8292(2)	7.8562(1)
V ($Å^3$)	240.29(1)	242.92(1)
La 4c (x,y,1/4)		
x	1.0011(3)	0.9996(3)
y	0.0249(2)	0.0246(1)
B_{iso}	1.1(2)	1.5(2)
f_{occ}	1.00	1.00
Ti/M 4b (1/2,0,0)		
B_{iso}	0.30	0.30
Ti/M f_{occ}	0.50/0.50	0.50/0.50
O1 4c (x,y,1/4)		
x	0.0678(4)	0.0661(3)
y	0.4992(3)	0.4906(2)
B_{iso}	1.4(4)	0.9(2)
f_{occ}	1.00	0.91(2)
O2 8d (x,y,z)		
x	0.7163(2)	0.7182(2)

y	0.2717(3)	0.2826(2)
z	0.0376(2)	0.0404(1)
$\mathbf{B_{iso}}$	1.2(2)	1.4(2)
$\mathbf{f_{occ}}$	1.00	1.00
Reability factors		
χ^2	0.748	1.02
$\mathbf{R_p}$ **(%)**	4.83	4.81
$\mathbf{R_{wp}}$ **(%)**	5.90	6.16
$\mathbf{R_{exp}}$ **(%)**	7.16	6.29
$\mathbf{R_B}$ **(%)**	9.54	9.84

Table II contains the interatomic distances and angles after the Rietveld refinements of both oxidized and reduced samples at room temperature. The (Co,Ti-O1) bond-lengths at room temperature for the oxidized phase compare reasonably well with the expected value calculated as ionic radii sums, displaying values of 1.9931(4) Å *versus* the calculated 2.026 Å ($^{VI}Co^{2+}$: 0.650 Å; $^{VI}Ti^{4+}$: 0.605 Å) [15]. As expected, the <(Co,Ti)-O2> distances for the reduced sample (2.006 Å) increase as a consequence of the increase of the ionic size from Ti^{4+} (i.r.: 0.605 Å) to Ti^{3+} (i.r.: 0.650 Å).

Table II. Selected atomic distances (Å) and angles (deg) for the oxidized and reduced $LaM_{0.5}Ti_{0.5}O_3$ (M= Co and Ni) at 295 K.

M	Co (oxid)	Co (red)
Distances (Å)		
La – O1	2.932(2)	2.993(2)
	2.651(2)	2.617(2)
	3.158(3)	3.151(2)
	2.395(3)	2.422(2)
<La – O1>	**2.784**	**2.795**
Ti/M – O1 (x2)	1.9931(4)	1.9988(3)
Ti/M – O2 (x2)	1.946(1)	2.010(1)
(x2)	2.039(1)	1.993(1)
<Ti/M – O2>	**1.993**	**2.006**
Angles		
Ti/M – O1 – Ti/M	158.25(2)	158.59(2)
Ti/M – O2 – Ti/M	158.84(5)	157.04(3)

3.2 Thermal analysis (TGA)

The thermal evolution of the sample was studied by recording TGA curves. Heating $LaCo_{0.5}Ti_{0.5}O_3$ in $5\%H_2/95\%N_2$ atmosphere leads to the reduction of the sample to give $LaCo_{0.5}Ti_{0.5}O_{3-\delta}$ with the same crystal structure. The left panel of Fig 4 depicts the stability of the oxidized sample; in the right panel the auxotherm run is followed by an isotherm treatment at 600°C required to stabilize the stoichiometry of the reduced sample, displaying the loss of 0.12 oxygen atoms at this temperature. The calculated value is in good agreement with the NPD data (Table I), leading to a $LaCo_{0.5}Ti_{0.5}O_{2.88}$ composition. The thermal analysis confirmed the existence of a mixed valence in the reduced phase. A thermal treatment of the resulting reduced phase in oxidizing (air) atmosphere restores the perovskite phase, thus confirming the required reversibility upon cycling in oxidizing-reducing atmospheres.

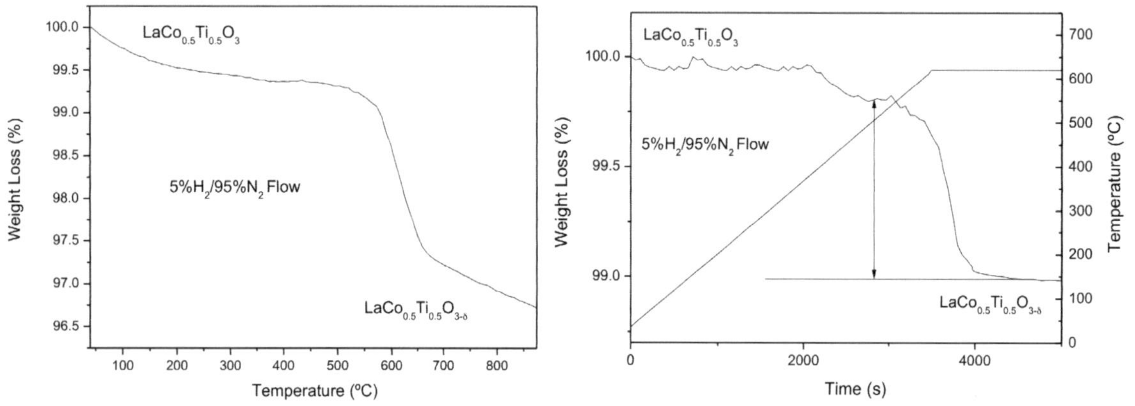

Fig. 4. Left panel: TG curve of $LaCo_{0.5}Ti_{0.5}O_3$ in reducing conditions; right panel: auxotherm run followed by an isotherm treatment at 600ºC where the reduced specimen is stabilized.

3.3. Thermal expansion measurements.

Aiming to determine the mechanical compatibility of our material with the other cell components, thermal expansion measurement of the dense ceramic was carried out in different atmospheres. The thermal expansion of the perovskite phases were measured in sintered pellets, initially heated in air at 900 ºC for 12 h; the reduced phase was finally treated in a 5% H_2 flow at 800 ºC for 4h. A dilatometric analysis was performed between 35 and 800 ºC for several cycles; the data where only recorded during the heating runs. Fig. 5 shows no abrupt changes in the thermal expansion of oxidized and reduced $LaCo_{0.5}Ti_{0.5}O_3$ in all the temperature range under measurement. The TEC measured in air atmosphere between 100 and 800 ºC is $10.43x10^{-6}$ K^{-1} for the oxidized phase. The thermal expansion of the reduced phase shows a value of $12.63x10^{-6}$ K^{-1} when heating the sample in $H_2(5\%)/N_2(95\%)$, very similar to that obtained for the oxidized perovskite. The determined TEC values match with those of other cell components and are certainly lower than those reported for some cobaltites, as expected by the partial substitution of Co by Ti.

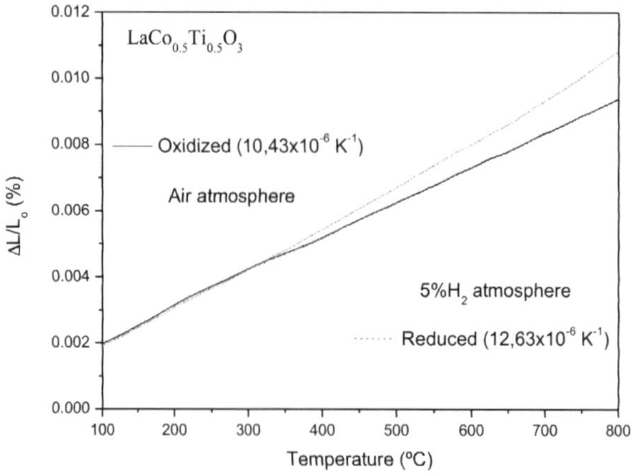

Fig.5. Thermal expansion determined by dilatometry for $LaM_{0.5}Ti_{0.5}O_3$ (oxidized and reduced).

3.4 Electrical conductivity measurements

Fig. 6 shows the thermal variation of the electrical conductivity of $LaCo_{0.5}Ti_{0.5}O_3$ measured twice, in order to be sure, in sintered bars in $5\%H_2/95\%N_2$ atmosphere by the *dc* four-probe method. The reduced phase shows a semiconductor-like behavior under reducing conditions with a maximum value of 0.27 $S \cdot cm^{-1}$ at 850 °C. Although the conductivity values obtained are lower than 0.5 Scm^{-1}, it is not necessary such high values for the anode materials in single oxide fuel cells. Although the reduced $LaCo_{0.5}Ti_{0.5}O_{3-\delta}$ perovskite was prepared under a $5\%H_2/95\%N_2$ flow at 600 °C, this phase is stable up to 750 °C, as checked by TG measurements. Beyond this temperature there is a slight weight loss which suggests a partial reduction of the oxygen stoichiometry; this would correspond only to the last two conductivity points, and, indeed, both conductivity and thermal expansion curves show a monotonic behavior in all the temperature range

Fig. 6 also illustrates the electrical conductivity of the oxidized $LaCo_{0.5}Ti_{0.5}O_3$ perovskite, measured in an air atmosphere. The phase display also a semiconductor-like behavior, with conductivity values much lower than for the reduced sample with a maximum value of 0.07 Scm^{-1} at 900°C. The reduced $LaCo_{0.5}Ti_{0.5}O_3$ perovskite, containing mixed Ti^{3+}-Ti^{4+} valence, presents a higher electrical conductivity by electron hopping between adjacent Ti^{3+} and Ti^{4+} cations (due to the antisite disordering), but the presence of M^{2+} ions hinders the long-range conduction path and promotes a low electronic conductivity.

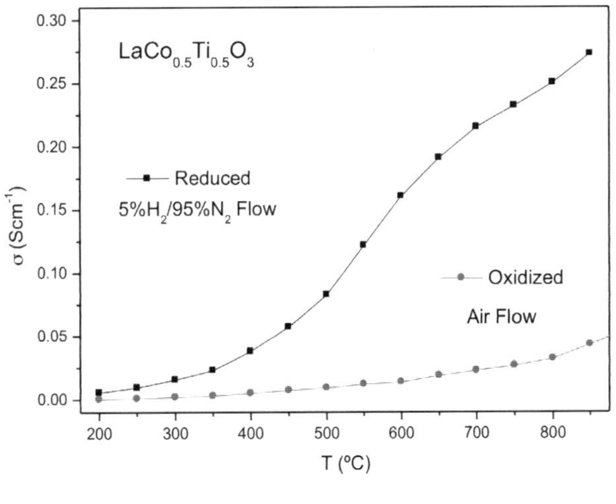

Fig.6. *dc*-conductivity as a function of temperature for $LaCo_{0.5}Ti_{0.5}O_3$ (oxidized and reduced).

4. Conclusions

In this work, we have prepared oxygen-stoichiometric $LaCo_{0.5}Ti_{0.5}O_3$ perovskite, containing Co^{2+} and Ti^{4+}, by soft chemistry procedures followed by thermal treatments in air. A topotactic reduction of the stoichiometric perovskites, in a reduced atmosphere, leads to oxygen-deficient phase with $LaCo_{0.5}Ti_{0.5}O_{2.91}$ compositions, where Ti^{4+} is partially reduced to Ti^{3+}, as shown by both neutron diffraction and thermogravimetric analysis. The expansion of the unit-cell volume ions in the reduced sample, the increase of the <Co,Ti-O> bond lengths and the localization of oxygen vacancies at the axial positions of the perovskite are sizeable proofs of the presence of Ti^{3+} cations in the specimen.

The crystal structure of both the oxidized and reduced $LaCo_{0.5}Ti_{0.5}O_3$ have been refined at RT in the orthorhombic *Pbnm* space group; the Co and Ti atoms are randomly distributed and no crystallographic long-range order was observed. The electrical characterization evidences a semiconductor behavior in all the samples displaying a maximum value of 0.27 Scm^{-1} in the reduced specimen. The thermal expansion coefficients for the oxidized and reduced phases perfectly match with the standard values of SOFCs electrolytes. The reversibility of the reduction-oxidation of the $LaCo_{0.5}Ti_{0.5}O_3$ makes it possible the required cyclability of the cells.

Acknowledgments
We thank the financial support of the Spanish Ministry of Science and Innovation (MICINN) to the project MAT2010-16404; AA and DPC also wish to thank to MICINN for a "Juan de la Cierva" and a "Ramón y Cajal" contracts, respectively. We are grateful to the Institut Laue-Langevin (ILL) in Grenoble for making all facilities available.

References

[1] Mitchell RH 2002 *Perovskites, Modern and Ancient* (Almaz Press, Ontario)
[2] Aguadero A, Alonso JA, Martínez-Lope MJ, Fernández-Díaz MT 2011 *Solid State Sciences* **13** 13-18.
[3] Tokura Y, Taguchi T, Okada Y, Fujishima Y, Arima T, Kumagai K, Iye Y. 1993 *Phys. Rev. Lett.* **70 (14)** 2126–2129.
[4] Okaka Y, Arima T, Tokura Y, Murayama C, Mori N. 1993 *Phys. Rev. B* **48 (13)** 9677–9683.
[5] Robey SW, Hudson LT, Eylem C, Eichorn B. 1993 *Phys. Rev. B* **48 (1)** 562–568.
[6] Sunstrom JE, Kauzlarich SM. 1993 *Chem. Mater.* **5** 1539–1544.
[7] Kamegashira N, Nakajima N, Watanabe K, Kobayashi M. 2000 *Journal of Alloys and Comp.* **311** 74-78.
[8] Alvarez-Serrano I, Cuello GJ, Lopez ML, Jimenez-Lopez A, Pico C, Rodriguez-Castellon E, Rodriguez E, Veiga ML. 2008 *J. Phys. D: Appl. Phys.* **41** 195001.
[9] Holman KL, Huang Q, Klimczuk T, Trzebiatowski K, Bos JWG, Morosan E, Lynn JW, Cava JR. 2007 *Journal of Solid State Chemistry.* **180** 75–83.
[10] Rodríguez E, López ML, Campo J, Veiga ML, Pico C. 2002 *Journal of Materials Chemistry* **12 (9)** 2798-2802.
[11] Cairns DL, Reaney IM, Zheng H, Iddles D, Price T. 2005 *Journal of European Ceramic Society.* **25** 433-439.
[12] Ramadass N, Gopalakrishnan J, Sastri MVC 1978 *J. Inorg. Nucl. Chem.* **40** 1453–1454
[13] Rietveld HM 1969 *J. Appl. Crystallogr.* **2** 65-71.
[14] Rodríguez-Carvajal J 1993 *Physica B.* **192** 55-69.
[15] Shannon RD 1976 *Acta Crystallogr. A* **32** 751-767.

High-pressure preparation and characterization of $(R,R')NiO_3$ (R,R'= rare earths) perovskites: effect of the variance of the R^{3+} ionic sizes

M J Martínez-Lope[1], J Sánchez-Benítez[2], M T Fernandez-Díaz[3] and J A Alonso[1]

[1]Instituto de Ciencia de Materiales de Madrid (CSIC), E-28049 Madrid, Spain
[2]Departamento de Química Física I, Fac. CC. Químicas, Universidad Complutense de Madrid, E-28040 Madrid, Spain
[3]Institut Laue Langevin, BP 156X, Grenoble, F-38042, France

E-mail: ja.alonso@icmm.csic.es

Abstract. In this paper we describe some novel members of the $RNiO_3$ perovskites, containing Ni^{3+} and presenting metal-insulator transitions coupled to a charge disproportionation effect. This phenomenon is observed in the insulating state as a subtle monoclinic distortion, due to the presence of two unequivalent Ni ions with different coordination environments, corresponding to expanded and contracted octahedra alternating in the three-dimensional network. Here we examine how the introduction of mixed rare-earth cations at the R sublattice affects the metal-insulator transition and the structural features, from high resolution neutron powder diffraction studies.

1. Introduction

High-pressure synthesis is well known to be an effective tool to stabilize transition-metal oxides in high-valence states (e.g. Mn(V), Fe(IV), Co(IV), Ni(III), Cu(III)) or intermediate-valence states (such as Cu(II)-Cu(III), Mn(III)-Mn(IV), Fe(III)-Fe(IV)) [1,2]. The stabilization of high oxidation states is achieved either under the high-pressure of an oxidizing gas, typically oxygen, or as a consequence of the compressive effect of the external pressure on the chemical bonds, favouring the shorter and more covalent bonds corresponding to the elevated valences. Happily, many solids synthesized at high pressure-high temperature conditions (where they are thermodynamically stable) can be "quenched" to ambient conditions, where they are termodynamically metaestable, yet they remain indefinitely kinetically stable. The heavy equipment usually required for high-pressure synthesis and the limited number of laboratories where this equipment is available justifies that many of these metastable oxides have only been superficially investigated, despite the exotic properties they may present such as superconductivity, metal behavior, metal-insulator transitions or colossal magnetoresistance. These phenomena are bound to the strong electronic correlations that they usually present.

In our research group we have paid much attention to a good number of metastable materials belonging to several families such as $RNiO_3$ (R= rare earths) perovskites [3-14], $Tl_2Mn_2O_7$ pyrochlore and derivatives [15-23], $RCu_3Mn_4O_{12}$ (R= rare earths) complex perovskite and derivatives [23-28], ABO_3 (A= Se, Te; B= divalent transition metals) perovskites containing p-block elements at A positions [29,30], etc. In the investigation of these oxides it has been decisive the use of neutron diffraction, most of the times in powder samples. This technique has allowed us to access to structural details typically related to the octahedral tilting in perovskite structures, the oxygen stoichiometry and order-disorder of the oxygen sublattice, the distinction between close elements in the Periodic Table, the resolution of magnetic structures and, in general, the establishment of a correlation between the structure and the properties of interest.

A central theme in our research has been the systematic study of the metal-to-insulator (MI) transitions in the RNiO$_3$ (R= rare earths) perovskites, that Torrance *et al.* [30] had interpreted as a function of the reduction and closing of the charge-transfer gap between O^{2-} and Ni^{3+} ions. The difficulty of preparation of these phases, associated with the presence of trivalent nickel, increases as the lanthanide radius diminishes, in such a way that the mentioned transport properties (and MI transitions) had only been investigated for R between La and Eu. Since the pioneering work by Démazeau in 1971 [31], the phases for R= Y, Gd, Tb, Dy, Ho, Er, Tm, Yb, and Lu, had not been prepared again. One of the most relevant results was the preparation of these materials under moderate pressures of 2 GPa, in closed gold capsules in the presence of $KClO_4$, in sufficiently large amounts to perform accurate neutron and synchrotron diffraction studies across the transitions. For $YNiO_3$ we described, for the first time, a **charge disproportionation** associated with the MI transition [4,5]. This effect is concomitant, in the insulator state, with an extremely subtle monoclinic distortion; in the crystal structure Ni atoms occupy two independent crystallographic sites with slightly different charge, 3+δ and 3-δ. This effect is extensive to other small rare-earth members, from R= Ho to R= Lu [6,7]. In this work we have prepared new members of this family of perovskites with mixed occupancy of the R positions by different rare-earths, mimicking the ionic radius of Er^{3+} and Y^{3+} found in the well-known $ErNiO_3$and $YNiO_3$ perovskites, in order to get information about the evolution of the crystal structure and metal-insulator transitions in these novel materials and how they are affected by the intrinsic disorder at the R site.

2. Experimental

The access at moderate pressures to the metastable RNiO$_3$ oxides requires the careful choice of the starting reactants. In order to synthesize the following four samples, $Nd_{0.09}Tm_{0.91}NiO_3$, $Nd_{0.32}Lu_{0.68}NiO_3$, $Y_{0.40}Tm_{0.60}NiO_3$, and $Y_{0.56}Yb_{0.44}NiO_3$, stoichiometric amounts of R_2O_3 (R= Nd, Tm, Yb, Y) and $Ni(OH)_2$ were mixed up and ground with $KClO_4$ (30 % in weight), introduced into gold capsules (8 mm diameter, 10 mm length), sealed and placed into a cylindrical graphite heater. The reactions were carried out in a piston–cylinder press (Rockland Research Co.), at a pressure of 2 GPa at 900 ºC for 20 min. The decomposition of $KClO_4$ provides an "in-situ" high O_2 pressure that is necessary to stabilize Ni^{3+} ions. The raw products, obtained as dense, homogeneous pellets, were ground and washed in water, in order to dissolve KCl coming from the decomposition of $KClO_4$; then the powder samples were dried in air at 150 ºC for 1 h. The final products were single phase to diffraction measurements.

The products were initially characterized by laboratory X-ray diffraction (XRD) (Cu Kα, λ= 1.5406 Å). Neutron Powder Diffraction (NPD) diagrams were collected at the Institut Laue-Langevin (ILL) in Grenoble (France). The crystal structures were refined from high-resolution NPD patterns at the D2B diffractometer with λ= 1.594 Å. The refinements of the crystal structures were performed by the Rietveld method [32], using the FULLPROF refinement program [33]. A pseudo-Voigt function was chosen to generate the line shape of the diffraction peaks. The following parameters were refined in the final runs: scale factor, background coefficients, zero-point error, pseudo-Voigt corrected for asymmetry parameters, positional coordinates and isotropic thermal factors.

Differential Scanning Calorimetry (DSC) measurements were performed in a Mettler TA3000 system equipped with a DSC30 unit, in the temperature range from 25 ºC to 590 ºC. The heating and cooling rates were 10 ºC·min^{-1}, using about 70 mg of sample in each run.

3. Results and Discussion

(R,R')NiO_3 samples were obtained as well crystallized powders. The X-ray patterns (Figure 1) are characteristic of monoclinic perovskites, space group $P2_1/n$, showing sharp and well-defined superstructure reflections due to the tilting of the NiO_6 octahedra. The unit-cell parameters are related to a_0 (ideal cubic perovskite, $a_0 \approx 3.8$ Å) as $a \approx \sqrt{2}a_0$, $b \approx \sqrt{2}a_0$, $c \approx 2a_0$.

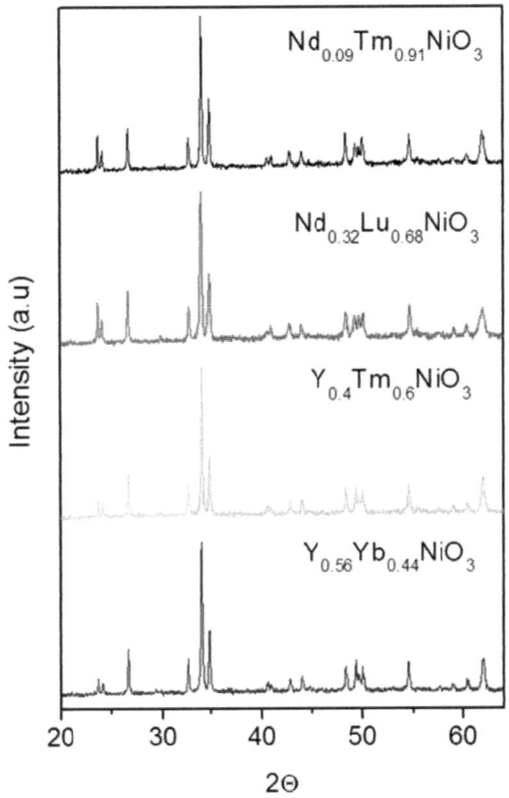

Figure 1. XRD patterns (Cu Kα, λ= 1.5406 Å) of *(R,R')*NiO_3 oxides.

A plot of the crystal structure is displayed in Figure 2.

The refinement of the crystal structures has been carried out from four NPD patterns acquired in the high-resolution diffractometer D2B at T= 295 K. The previously reported RT structures for *R*NiO_3 (*R*= Ho, Y, Er and Lu) [6] were used as starting model. In the $P2_1/n$ phase there are two crystallographically independent Ni positions (Ni1 and Ni2), as well as three kinds of non-equivalent oxygen atoms (O1, O2 and O3) all in general positions. $Ni1O_6$ and $Ni2O_6$ octahedra alternate along the three directions of the structure, in such a way that each $Ni1O_6$ octahedron is linked to six $Ni2O_6$ octahedra, and vice-versa. A minor NiO impurity was included as a second phase during the refinement, and vanadium metal, coming from the double-walled sample holder, as a third phase. Figure 3 illustrates the goodness of the structural refinement from NPD data at RT for $Nd_{0.09}Tm_{0.91}NiO_3$, $Nd_{0.32}Lu_{0.68}NiO_3$, $Y_{0.40}Tm_{0.60}NiO_3$, $Y_{0.56}Yb_{0.44}NiO_3$. Table 1 contains the structural parameters after the refinements. The main bond distances and bond angles are listed in Table 2.

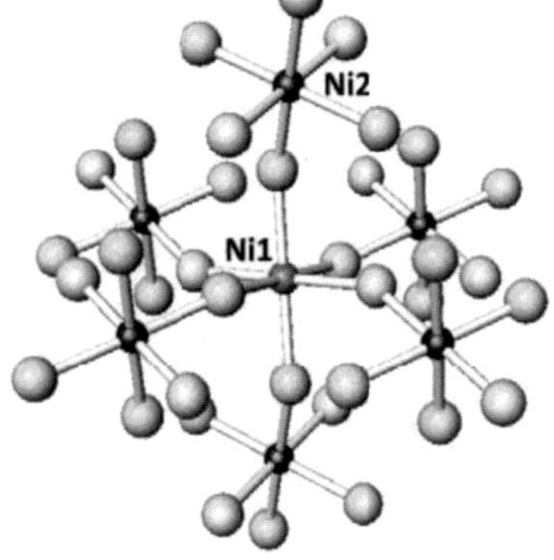

Figure 2. View of the crystal structure of $RNiO_3$ (R= rare earths) in the insulating monoclinic region below T_{MI}. Corner-sharing NiO_6 octahedra are fairly tilted in the structure to optimize R-O bond-lengths; two kind of NiO_6 octahedra (expanded and compressed) alternate in the 3D network.

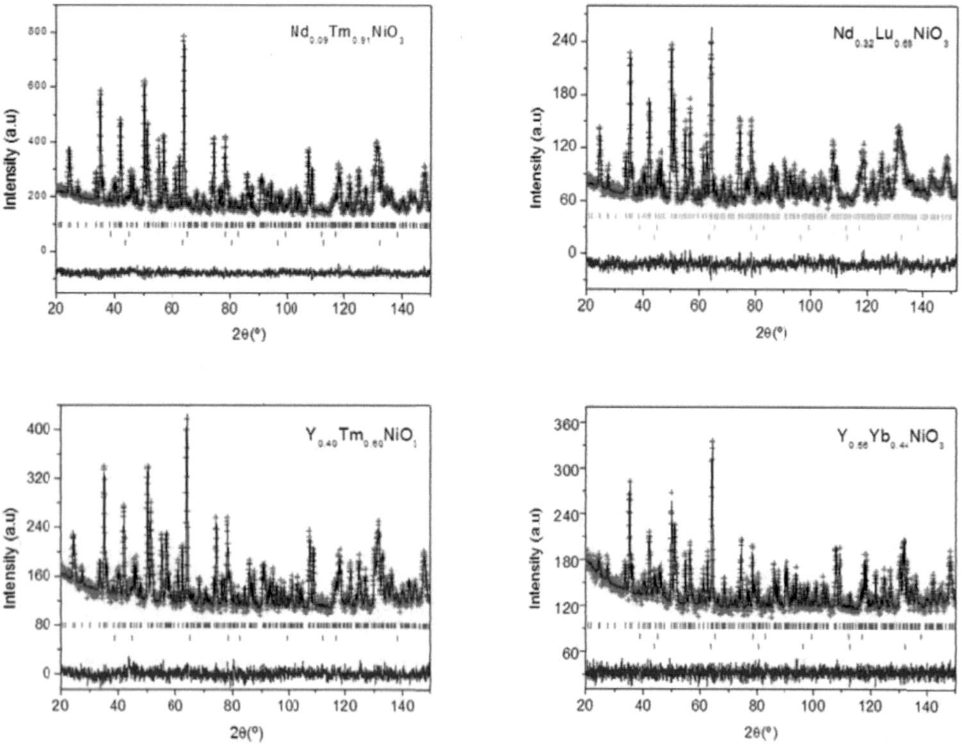

Figure 3. Observed (crosses), calculated (solid line) and difference (bottom) NPD Rietveld profiles (λ= 1.594 Å) for $Nd_{0.09}Tm_{0.91}NiO_3$, $Nd_{0.32}Lu_{0.68}NiO_3$, $Y_{0.40}Tm_{0.60}NiO_3$, and $Y_{0.56}Yb_{0.44}NiO_3$ at RT. Second and third line of Bragg reflections corresponds to the NiO impurity and vanadium holder, respectively.

Table 1. Unit-cell, positional, thermal parameters and reliability factors after the refinements of the crystal structure of (*R*,*R'*)NiO_3 (*R*= Nd, Y; *R'*= Tm, Yb, Lu) perovskites in the monoclinic $P2_1/n$ space group, Z= 4, from NPD data at 295 K. $<r_R>$ is the average ionic size of R and R' and σ^2 the variance.

(*R*,*R'*)NiO_3		$Nd_{0.09}Tm_{0.91}$	$Nd_{0.32}Lu_{0.68}$	$Y_{0.40}Tm_{0.60}$	$Y_{0.56}Yb_{0.44}$
T_{MI} heating (° C)		304.9	287.4	312.9	313.2
T_{MI} cooling (° C)		302.9	284.6	309.8	309.9
$<r_R>$		1.004	1.019	1.004	1.004
σ^2		0.00108	0.00379	0.00015	0.00029
a(Å)		5.1618(1)	5.1608(1)	5.1574(2)	5.1563(2)
b (Å)		5.5004(1)	5.4934(3)	5.5079(2)	5.5099(2)
c (Å)		7.3947(2)	7.4006(4)	7.3898(2)	7.3881(3)
β		90.122(4)	90.232(6)	90.106(4)	90.104(4)
V (Å^3)		209.95(9)	209.81(2)	209.92(1)	209.90(1)
R	4e (x y z)				
x		0.9785(5)	0.9770(8)	0.9804(9)	0.981(1)
y		0.0737(3)	0.0723(4)	0.0753(4)	0.0754(6)
z		0.249(1)	0.252(1)	0.250(2)	0.253(2)
B(Å^2)		0.37(3)	0.35(6)	0.48(5)	0.48(8)
Ni1	2d (½ 0 0)				
B(Å^2)		0.58(8)	0.03(9)	0.6(1)	0.4(2)
Ni2	2c (½ 0 ½)				
B(Å^2)		0.30(8)	1,01(9)	0.5(1)	0.5(2)
O1	4e (x y z)				
x		0.1031(5)	0.1040(9)	0.1019(8)	0.102(1)
y		0.4689(5)	0.4671(8)	0.4665(8)	0.467(1)
z		0.2446(9)	0.249(1)	0.244(1)	0.2472)
B(Å^2)		0.59(5)	0.9788)	0.52(7)	0.6(1)
O2	4e (x y z)				
x		0.7025(9)	0.697(2)	0.701(2)	0.698(3)
y		0.308(1)	0.310(2)	0.311(2)	0.313(3)
z		0.0503(8)	0.053(1)	0.051(1)	0.051(2)
B(Å^2)		0.44(8)	0.77(13)	0.58(13)	0.7(2)
O3	4e (x y z)				
x		0.1821(9)	0.191(1)	0.182(1)	0.186(2)
y		0.202(1)	0.203(2)	0.204(2)	0.203(2)
z		0.946288)	0.950(1)	0.947(1)	0.947(2)
B(Å^2)		0.73(9)	0.60(13)	0.54(13)	0.4(2)
Reliability factors					
χ^2		1.16	1.43	1.07	0.927
R_p(%)		2.99	3.97	3.43	3.22
R_{wp}(%)		3.75	5.01	4.35	4.03
R_{exp}(%)		3.47	4.19	4.20	4.19
R_I(%)		4.11	5.64	5.92	4.63

The occurrence of a MI transition in the $RNiO_3$ perovskites was checked by DSC measurements. Figure 4 illustrates the DSC curves obtained during the heating and cooling runs for $Y_{0.40}Tm_{0.60}NiO_3$. The heating process exhibits an endothermic peak centered at 313 ºC. The reverse transition, showing an exothermic peak, is observed during the cooling run at 310 ºC. By analogy with the DSC peaks observed for other rare-earth nickelates [6], the peaks shown in figure 4 can be assigned to the corresponding MI transition. The heat transfer accounts for the entropy and lattice energy gain due to the electronic delocalization. The T_{MI} temperatures of the four samples are gathered in Table 1.

Table 2. Main bond distances (Å) and selected angles (deg) for monoclinic (*RR'*)NiO_3 (*R*= Nd, Y; *R'*= Tm, Yb, Lu) perovskites at room temperature.

(*R*,*R'*)	$Nd_{0.09}Tm_{0.91}$	$Nd_{0.32}Lu_{0.68}$	$Y_{0.40}Tm_{0.60}$	$Y_{0.56}Yb_{0.44}$
Ni1 octahedra				
Ni1 – O1 (x2)	1.971(6)	1.946(9)	1.972(9)	1.947(14)
Ni1 – O2 (x2)	2.023(6)	2.023(9)	2.039(9)	2.039(16)
Ni1 – O3 (x2)	2.021(5)	1.980(8)	2.025(7)	2.005(23)
<Ni1 – O>	2.005(6)	1.983(9)	2.012(8)	1.997(14)
Δ_d 10^4	1.44	2.52	2.06	3.62
Ni2 octahedra				
Ni2 – O1 (x2)	1.892(6)	1.924(9)	1.888(9)	1.910(14)
Ni2 – O2 (x2)	1.902(5)	1.921(9)	1.896(8)	1.906(16)
Ni2 – O3 (x2)	1.930(6)	1.943(8)	1.920(8)	1.937(15)
<Ni2 – O>	1.908(6)	1.929(9)	1.901(8)	1.921(15)
Δ_d 10^4	0.71	0.26	0.51	0.87
<Ni – O>	1.956(6)	1.956(9)	1.956(8)	1.959(14)
Ni1– O1 – Ni2	146.3(3)	146.0(4)	146.4(4)	146.6(6)
Ni1 – O2 – Ni2	147.8(2)	145.7(4)	146.9(3)	146.1(7)
Ni1 – O3 – Ni2	145.3(2)	147.7(3)	146.0(3)	146.3(6)
RO_9 polyhedra				
R – O1	2.26783)	2.266(5)	2.244	2.248(8)
R – O1	3.057(4)	3.054(6)	3.063(6)	3.064(9)
R – O1	2.236(4)	2.238(6)	2.236(6)	2.231(9)
R – O2	2.418(8)	2.44(1)	2.434(1)	2.463(18)
R – O2	2.283(8)	2.23(1)	2.27(1)	2.246(18)
R – O2	2.588(9)	2.58(1)	2.57(1)	2.543(19)
R – O3	2.578(9)	2.60(1)	2.57(1)	2.592(19)
R – O3	2.255(8)	2.29(1)	2.281(1)	2.298(18)
R – O3	2.446(8)	2.42(1)	2.44(1)	2.424(16)
<R – O>$_{8short}$	2.384(8)	2.38(1)	2.380(9)	2.381(16)
<R – O>	2.720(8)	2.72(1)	2.720(9)	2.724(16)

For all the studied *R*NiO_3 perovskites the MI transition occurs considerably above room temperature in such a way that at RT the structural data correspond to samples with a 100% of insulating fraction. The structural refinement in the monoclinic symmetry is unique and shows a crystallographic description completely comparable to those obtained for the smaller rare-earth *R*NiO_3 compounds (*R*= Ho, Er, Tm, Yb and Lu). In the insulating regime below T_{MI}, Ni1 and Ni2 sites alternate along the three crystallographic directions (Figure 2). As shown in Table 2, the mean Ni-O distance in the $Ni1O_6$ octahedron (e.g. 2.012(8) Å in $Y_{0.40}Tm_{0.60}NiO_3$) is considerably larger than in the $Ni2O_6$ polyhedron (1.901(8) Å) at RT; we obtain again the same picture of expanded and contracted NiO_6 octahedra alternating along the three axes (Figure 2), consequence of the charge disproportionation phenomenon in the insulating regime. The difference between both mean Ni-O distances is slightly higher (within the standard deviations) to that observed for the smaller rare earth nickelates (*R* = Y, Ho, Er, Lu) [7,8]. It is remarkable that the distortion of the $Ni1O_6$ octahedra is much more pronounced than that of $Ni2O_6$ octahedra; at RT the distortion parameter for Ni1, defined as $\Delta_d = (1/6)\Sigma_{n=1,6}[(d_n-<d>)/<d>]^2$ is 3.62 10^{-4} for $Y_{0.56}Yb_{0.44}NiO_3$, whilst for Ni2 it is 0.87 10^{-4}. The observed static distortion is even higher than those found for the heaviest rare earths, falling in the interval Δ_d=0.8-1.7·10^{-4} [7,8].

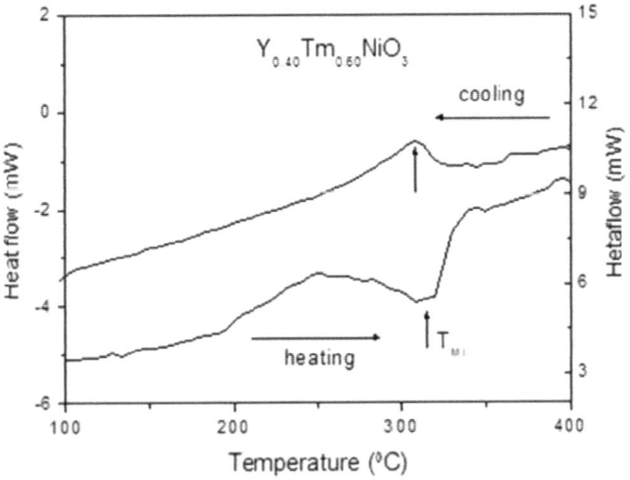

Figure 4. DSC curves for $Y_{0.40}Tm_{0.60}NiO_3$ obtained on the heating and cooling runs.

The proportions of *R* and *R*' in (*RR*')NiO_3 were chosen in such a way that three of them ($Nd_{0.09}Tm_{0.91}$, $Y_{0.40}Tm_{0.60}$ and $Y_{0.56}Yb_{0.44}$) mimic the ionic size of Er^{3+} (i_r= 1.004 Å) and one of them ($Nd_{0.32}Lu_{0.68}$) that of Y^{3+} (i_r= 1.019 Å). These substitutions usually result in size disorder or mismatch. This effect can be quantified by the variance of the A-site cation distribution, defined as $\sigma^2=<r_R^2>-<r_R>^2$ (r_R= ionic radius of R cation). Table 1 contains the mean ionic R radius and the variance calculated for all the samples. Figure 5 shows the linear dependence of the T_{MI} with the variance. It is interesting to see that the compounds including Nd have a stronger deviation from the MI temperatures of $ErNiO_3$ (311 ºC) and $YNiO_3$ (309 ºC), as a result of the larger variance of the ionic sizes, involving large (Nd^{3+}) and small (Tm^{3+} and Lu^{3+}) ions, whereas the Y-containing compounds, with a smaller variance of ionic radius, show a much smaller deviation of the transition temperature with respect to the unmixed $ErNiO_3$ and $YNiO_3$ perovskites. In fact, the metal-transition temperature scales with the variance and not with the mean ionic radius of the R cations as shown in Fig. 5. A similar effect was described for manganese perovskites with colossal magnetorresistance [34], for which a sizeable decrease of ferromagnetic T_C upon disordering at the A positions was observed, and interpreted as coming from strain fields resulting from ordered or disordered oxygen displacements that are parameterized by the statistical mean and variance of the A cation radius. An analogous interpretation can be given here; the suppression of the MI transition observed when large and small rare earth cations randomly occupy the A sites is a result of oxygen displacements inducing an irregular Ni-O bond-length and Ni-O-Ni angle distribution across the crystal, which narrows the bandwidth of the conduction band and diminishes T_{MI}.

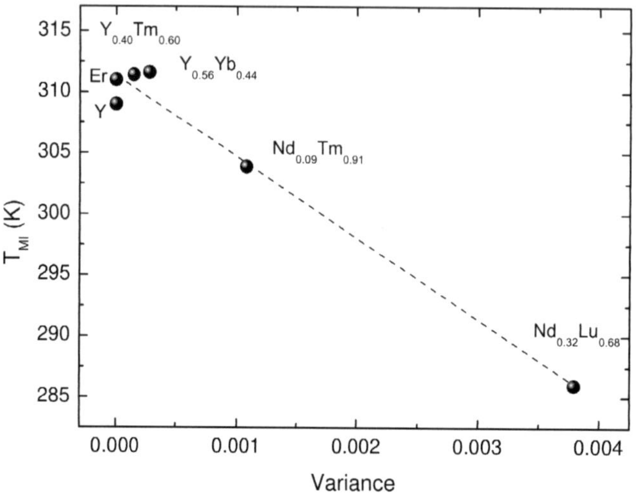

Figure 5. Variation of the T_{MI} transition with the variance of the four studied samples. The line is guide to the eye.

4. Conclusions

Highly distorted (RR')NiO_3 perovskites, containing different rare-earth cations (R= Nd, Y, Tm, Yb) with diverse variance were prepared under high-oxygen pressure. The metal-insulator transition temperatures were determined by DSC measurements; they occur well above RT, with T_{MI} ranging between 284 and 310 ºC. The crystal structure at room temperature, studied from high-resolution neutron powder diffraction, shows that these oxides stabilize a charge disproportionation below T_{MI} implying a separation of two kinds of (NiO_6) octahedra (large $Ni1O_6$ and small $Ni2O_6$) described in a pseudo-orthorhombic unit cell, characterized by a subtle monoclinic distortion (β between 90.23° and 90.10°) in the $P2_1/n$ space group. Whereas ($Ni1O_6$) octahedron is significantly distorted, ($Ni2O_6$) is much more regular. This is consistent with a scenario considering that the former contains more ionic Ni-O bonds and it is susceptible to experience a Jahn-Teller deformation, whereas the small ($Ni2O_6$) octahedron is constituted by more covalent Ni-O bonds. The analysis of the variance of the rare-earth distribution at the A positions shows that those perovskites containing rare earths with very different ionic sizes (large variance) exhibit a substantial reduction of the MI transition temperature, as a consequence of the irregular Ni-O bond-length and Ni-O-Ni angle distribution across the crystal, which narrows the bandwidth of the conduction band.

Acknowledgments

We thank the financial support of the Spanish Ministry of Science and Technology to the project MAT2010-16404 and of the Comunidad de Madrid to the project S2009PPQ-1551. We are grateful to the Institut Laue-Langevin (ILL) for making all facilities available.

References

[1] Goodenough J B, Kafalas J A and Longo J M 1972 *Preparative Methods in Solid State Chemistry* Ed. P. Hagenmuller (Academic Press) 2-69

[2] Demazeau G 1988 *Chim. Scripta* **28(1)** 21-24; Demazeau G, Fabritchnyl P B, Legasov V A, Etourneau J and Hagenmuller P 1989 *Russian J. Inorg Chem.* **34(1)** 1-6

[3] Alonso J A, Martínez-Lope M J and Rasines I 1995 *J. Solid State Chem.* **120** 170

[4] Alonso J A, García-Muñoz J L, Fernández-Díaz M T, Aranda M A G, Martínez-Lope M J and Casais M T 1999 *Phys. Rev. Lett.* **82** 3871

[5] Alonso J A, Martínez-Lope M J, Casais M T, Aranda M A G and Fernández-Díaz M T 1999 *J. Am. Chem. Soc.* **121** 4754

[6] Alonso J A, Martínez-Lope M J, Casais M T, García-Muñoz J L and Fernández-Díaz M T 2000 *Phys. Rev. B* **61** 1756

[7] Alonso J A, Martínez-Lope M J, Casais M T, García-Muñoz J L, Fernández-Díaz M T and Aranda M A G 2001 *Phys. Rev. B* **64** 94102

[8] Fernández-Díaz M T, Alonso J A, Martínez-Lope M J, Casais M T and García-Muñoz J L 2001 *Phys. Rev. B* **64** 144417

[9] De la Cruz F, Piamonteze C, Massa N E, Salva H, Alonso J A, Martínez-Lope M J and Casais M T 2002 *Phys. Rev. B* **66** 153104

[10] Lorenzo J E, Hodeau J L, Paolasini L, Lefloch S, Alonso J A and Demazeau G 2005 *Phys. Rev. B* **71** 45128

[11] Piamonteze C, Tolentino H C N, Ramos A Y, Massa N E, Alonso J A, Martínez-Lope M J and Casais M T 2005 *Phys. Rev. B* **71** 12104

[12] Presniakov I A, Baranov A V, Demazeau G, Rusakov V S, Sobolev A V, Alonso J A, Martínez-Lope M J and Pokholok K 2007 *J. Phys.: Condens. Matter* **19** 036201

[13] Caytuero A, Micklitz H, Abd-Elmeguid M M, Litterst F J, Alonso J A and Baggio-Saitovitch A 2007 *Phys. Rev. B* **76** 193105

[14] Alonso J A, Martínez-Lope M J, Casais M T, Martínez J L, Demazeau G, Largeteau A, García-Muñoz J L, Muñoz A and Fernández-Díaz M T 1999 *Chem. Mat* **11** 2463

[15] Alonso J A, Martínez J L, Martínez-Lope M J, Casais M T and Fernández-Díaz M T 1999 *Phys. Rev. Lett.* **82** 189

[16] Alonso J A, Velasco P, Martínez-Lope M J, Casais M T, Martínez J L, Fernández-Díaz M T and de Paoli J M 2000 *Appl. Phys. Lett.* **76** 3274

[17] Alonso J A, Martínez-Lope M J, Casais M T, Velasco P, Martínez J L, Fernández-Díaz M T and de Paoli J M 1999 *Phys. Rev. B* **60** R15024

[18] Venkatesan M, Velasco P, Alonso J A, Martínez J L and Coey J M D 2004 *J. Phys.: Cond. Matter* **16** 3465

[19] Massa N E, Alonso J A, Martínez-Lope M J, Casais M T and Salva H 1999 *Phys. Rev. B* **60** 7445; Velasco P, Alonso J A, Martínez-Lope M J, Casais M T, Martínez J L, Fernández-Díaz M T and de Paoli J M 2001 *Phys. Rev. B* **64** 184436

[20] Velasco P, Alonso J A, Martínez-Lope M J, Casais M T, Martínez J L, Fernández-Díaz M T and de Paoli J M 2001 *J. Phys: Condens. Mat.* **13** 10991

[21] Velasco P, Alonso J A, Tissen V G, Marshall W G, Casais M T, Martínez-Lope M J, de Andrés A, Prieto C and Martínez J L 2003 *Phys. Rev. B* **67** 104403

[22] Sánchez-Benítez J, Alonso J A, Martínez-Lope M J, Casais M T, Martínez J L, de Andrés A and Fernández-Díaz M T 2000 *Chem. Mater.* **15** 219

[23] Alonso J A, Sánchez-Benítez J, de Andrés A, Martínez-Lope M J, Casais M T and Martínez J L 2003 *Appl. Phys. Lett.* **83** 2623

[24] Falcón H, Sánchez-Benítez J, Martínez-Lope M J, Alonso J A, Krezhov K, Spirov I and Ruskov T 2007 *J. Phys.: Condens. Matter* **19** 356209

[25] Sánchez-Benítez J, Retuerto M, Martínez-Lope M J, de la Calle C, Fernández-Díaz M T and Alonso J A 2010 *Zeitschrift fur Kristallographie* **225** 201
[26] Sánchez-Benítez J, Alonso J A, Martínez-Lope M J, de Andrés A and Fernández-Díaz M T 2010 *Inorganic Chemistry* **49** 5679
[27] Sánchez-Benítez J, de Andrés A, García-Hernández M, Alonso J A and Martínez-Lope M J 2006 *Mater. Sci. Eng. B* **126** 262
[28] Muñoz A, Alonso J A, Martínez-Lope M J, Morán E and Escamilla R 2006 *Phys. Rev. B* **73** 104442
[29] Muñoz A, Alonso J A, Martínez-Lope M J, Falcón H, García-Hernández M and Morán E 2006 *Dalton Trans.* 4936-43
[30] Torrance J B, Lacorre P, Nazzal A I, Ansaldo E J and Niedermayer Ch 1992 *Phys. Rev. B* **45** 8209
[31] Demazeau G, Marbeuf A, Pouchard M and Hagenmuller P J 1971 *Solid State Chem.* **3** 582
[32] Rietveld H M 1969 *J. Appl. Crystallogr.* **2** 65
[33] Rodríguez-Carvajal J 1993 *Physica B* **192** 55
[34] Rodríguez-Martínez L M, Attfield J P 1996 *Phys. Rev. B* **54** R15622

Thermal expansion of $CaIrO_3$ post-perovskite determined by time-of-flight measurements

F. Aguado,[1] S. Hirai,[2] S. A. T. Redfern,[3] R. I. Smith[4]

[1]MALTA-Consolider team. Unidad Asociada ICMA-CSIC. DCITIMAC, Universidad de Cantabria, 39005-Cantabria, Spain
[2]Department of Geological & Environmental Sciences, Stanford University, Stanford CA 94305-2115, USA
[3]Department of Earth Sciences, University of Cambridge, Cambridge CB2 3EQ, UK
[4]Rutherford Appleton Lab, ISIS facility, Didcot OX11 0QX , UK

Email:aguadof@unican.es

Abstract. The thermal evolution of the $CaIrO_3$ post-perovskite structure has been determined by neutron powder diffraction (time-of-flight) measurements in a wide temperature range, from 1.8K to 550K. The linear expansion is similar to that found in previous x-ray diffraction studies, being $\alpha_b > \alpha_c > \alpha_a$. However, the difference in relative lattice parameters is less pronounced in the present case, suggesting a more isotropic evolution under temperature. Other structural differences found at low and high temperatures through several X-ray and neutron diffraction studies have been also analysed.

1. Introduction

Recent studies on different perovskites (*pv*) constituting the Earth's mantle revealed a new denser structure under extreme conditions.[1] This arrangement found in $MgSiO_3$ at high pressures and temperatures, i.e. the post-perovskite (*ppv*) phase, is the same as that seen in $CaIrO_3$ at room conditions: a layered structure of IrO_6 octahedra sharing corners and edges along different crystallographic directions within the plane (see fig. 1).[2] Accordingly, $CaIrO_3$ was initially proposed to carry out studies on compression at moderate P/T, being a good analogue for understanding the structural behaviour of the unquenchable $MgSiO_3$ *ppv* phase.[3] However, later reports challenged the suitability of $CaIrO_3$ to describe elastic properties of silicates at lower mantle conditions.[4] In addition, recent studies on isostructural compounds show different compression and expansion trends depending on the system.[5] Consequently, there is still an intense activity devoted to explain the deformation mechanisms found in this kind of compounds, which is particularly interesting to rationalize the structural behaviour of the ABX_3 family. Since $ASiO_3$ compounds are the main mantle constituents, it is especially important to clarify their behaviour at the mantle-core interface, i.e. the D'' layer, crucial to understand the seismic features found at that level.

$CaIrO_3$ can be isolated in either the *pv* or the *ppv* phase, just modifying the synthesis conditions. The *ppv* phase has been proved to be stable in wide pressure and temperature ranges.[6,7] However, concerning temperature solely, significant differences can be found in structures from different reports.[7-9] In addition, a full structural characterization in a wide temperature range hasn't been performed yet. Therefore, before analysing complex high P/T behaviours, it is important to properly determine the temperature effect on $CaIrO_3$ *ppv* structure.

Here we report the structural evolution of $CaIrO_3$-*ppv* in the 1.8-550 K range determined from time-of-flight (TOF) measurements on powder samples synthesized at high P/T. Thermal expansion results will be compared to available data from XRD and neutron diffraction experiments on samples obtained from different methods.

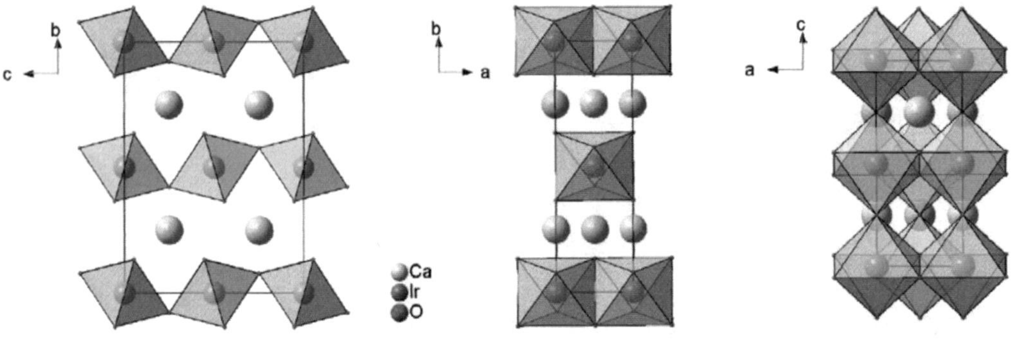

Figure 1 Views of the layered $CaIrO_3$ post-perovskite structure along the *a, c* and *b* axes.

2. Experimental

$CaIrO_3$ powder samples were obtained through high P/T synthesis by using a piston-cylinder apparatus. Stoichiometric amounts of starting oxides (IrO_2 and CO) were thermally pretreated and subsequently loaded in a graphite furnace-type assembly with a talc-pyrex outer sleeve. Different applied pressures (0.9-2.5 GPa), temperatures (1275K-1675K) and treatment durations (30 min-2hrs.) were explored, determining the optimal conditions to avoid Ca_2IrO_4 and $CaIrO_3$-*pv* in the final product. Thus, quenched samples showed no significant traces of such phases through XRD analysis. TOF neutron powder diffraction experiments (NPD) were carried out at the POLARIS instrument, ISIS facility. Given the limited sample amount and the strong iridium absorption, some pellets were loaded in a low signal vanadium can and placed in a standard cryofurnace. Several patterns were recorded in the 1.8K- 550K range, with counting times ranging from 7 h to 8h. Prior to any analysis, data were corrected for absorption to get appropriate intensities for full structural characterization. Subsequently, data from the highest resolution detector were used to perform full Rietveld refinements by means of the GSAS package.[10]

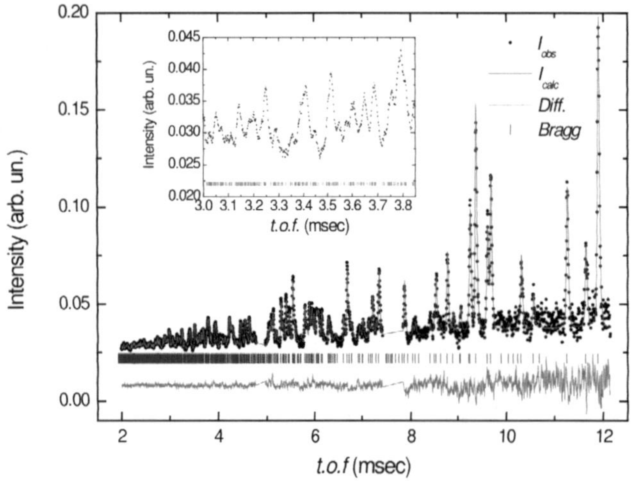

T(K):	295(1)
S.G.:	*Cmcm*
a (Å):	3.1412(1)
b (Å):	9.8782(4)
c (Å):	7.3040(3)
y_{Ca} (4c):	0.2500(3)
U_{Ca} (100·Å^2):	0.473
U_{Ir} (100·Å^2):	0.230
y_{O1}(4c):	0.4247(2)
U_{O1} (100·Å^2):	0.425
y_{O2}(8f):	0.1266(1)
z_{O2}(8f):	0.0505(2)
U_{O2} (100·Å^2):	0.362
R_{wp}:	0.025

Figure 2 Results of Rietveld refinement of $CaIrO_3$ at T=295 K. Note that other $a_i = 0$ except $z_{O1} = 1/4$.

3. Results and discussion

Rietveld refinements were carried out using the centrosymmetric Space Group *Cmcm*, which provided the best agreement factors. Thus, the use of the alternative non-centrosymmetric $Cmc2_1$ group is not justified.[2,9] Apart from the crystal *ppv* phase, no additional contributions (impurities or magnetic phases) were detected in any of the patterns within the experimental error. Figure 2 shows the results of the structural analysis at ambient temperature. Refined lattice parameters, the four non-special atomic positions and the isotropic temperature factors are included in the table. The structure is compatible with previous diffraction experiments.[2,7-9] It can be rationalized as a series of layers along the crystallographic *b* direction, formed by interconnected IrO_6 octahedra, sharing edges and corners along the *a* and *c* directions, respectively (see fig. 1). The Ir-O bond lengths (1.970Å and 2.040Å) and the Ir-O1-Ir tilting angle (θ = 135.8°) are very close to available values on single crystal and powder samples from ambient pressure synthesis.[2]

Table 1 includes a summary of results on the $CaIrO_3$ structure at selected low and high temperatures, comparing lattice from different reports (NPD and XRD). Thus, differences in lattice parameters and volumes within the same range are below 0.3% and 0.1%, respectively, whereas in the unit cell ratios are lower than 0.5% (the biggest discrepancies corresponding to RT). However, in spite of the small differences between them, it is necessary to carefully analyse their temperature evolution.

Table 1. Lattice parameters and unit cell metrics determined at selected temperatures for $CaIrO_3$.

	≈ 2 K		RT			≈ 550K		
	This study	Ref. 9	This study	Ref. 9	Ref. 7	This study	Ref. 6	Ref. 7
	NPD	NPD	NPD	NPD	XRD	NPD	XRD	XRD
a (Å)	3.1374	3.1345	3.1406	3.1365	3.1457	3.1466	3.1447	3.1499
b (Å)	9.8631	9.8757	9.8764	9.8835	9.8640	9.9000	9.9055	9.8932
c (Å)	7.2948	7.2952	7.3028	7.2990	7.2975	7.3183	7.3147	7.3148
V ($Å^3$)	225.73	225.83	226.52	226.26	226.44	227.97	227.95	227.95
b/a	3.144	3.151	3.145	3.151	3.136	3.146	3.150	3.141
c/a	2.325	2.327	2.325	2.327	2.320	2.326	2.326	2.322
T (K)	1.8	2.0	295	293	298	550	536	553

Figure 3 shows the relative lattice parameters and cell volume as a function of temperature in the 1.8- 550 K range. Previous data from Lindsay-Scott *et al.* (XRD)[7] have been included for comparison purposes. It can be observed that the expansion is anisotropic in both cases, although differences are much less pronounced in the present study. In both cases the *b* axis shows the greatest variation, that is, the interlayer expands more than any of the intralayer distances. In addition, *a* and *c* relative variations are quite similar, being the former (the edge-sharing direction) slightly smaller. Thus, results from thermal expansion indicate that $\alpha_b > \alpha_c > \alpha_a$. Interestingly, the pressure effect on lattice anisotropy is not exactly the opposite of thermal expansion, as one could expect: The *a* axis compresses less than the *c* axis, being $\beta_c > \beta_a > \beta_b$.[8,11] However their relative variation is quite similar, representing only 0.1% between 0 and 8 GPa, whereas Δb is approximately 1% over the *a* and *c* average in the same pressure range. This result agrees well with the rather isotropic expansion observed in the *a-c* plane through the present study. On the contrary, the isostructural $MgSiO_3$ presents

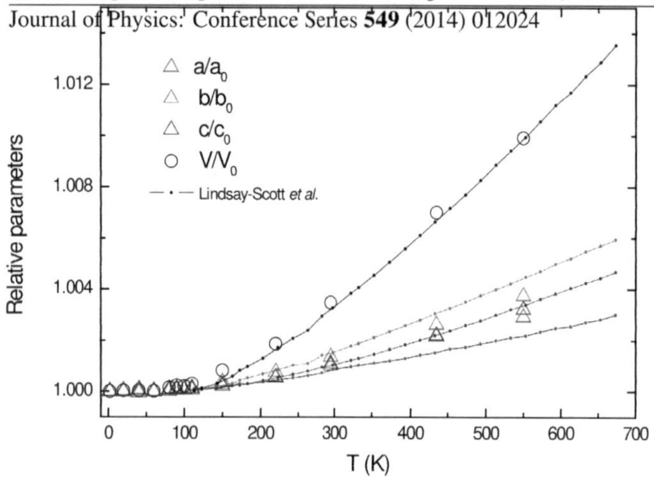

Figure 3. Temperature dependence of the relative volume and lattice parameters. Data from ref. 7 is also included for comparison.

a regular behaviour in this respect, showing α_i and $1/\beta_i$ the same sequence.[12] Nevertheless, the greatest expansion still corresponds to the interlayer direction ($\alpha_b > \alpha_a > \alpha_c$), unlike some results observed in the $CaBO_3$ family, where more differences can be found among *ppv* analogues. Thus, the expansion in $CaPtO_3$ differs from that of $CaIrO_3$ and $MgSiO_3$, being $\alpha_c > \alpha_b > \alpha_a$, whereas on compression the anisotropy is the same: $\beta_c > \beta_a > \beta_b$.[5] In $CaRhO_3$ the variation of the *c* axis is also the greatest, $\alpha_c > \alpha_a > \alpha_b$, decreasing the interlayer distance (*b* axis) monotonically above the magnetic phase transition (T_N= 90K).[13] Finally, $CaRuO_3$ presents a slight reduction of the *a* axis as temperature increases, as well as the same expansion sequence as in $CaIrO_3$: $\alpha_b > \alpha_c > \alpha_a$.[14] This plethora of thermal behaviours among analogues foresees a complex explanation of expansion systematics within the *ppv* family.

4. Conclusions

The thermal evolution of $CaIrO_3$ *ppv* structure has been established through a TOF experiment conducted at the ISIS facility. In the 1.8-550K range the *Cmcm* structure is stable and no additional phases were detected below the magnetic ordering temperature (T_C = 108K).[2] The thermal expansion is similar to that reported in literature, with $\alpha_b > \alpha_c > \alpha_a$, although significant quantitative differences have been found, being the expansion considerably more isotropic than previously reported. Nevertheless, discrepancies could be associated to sample synthesis procedure and additional experiments are required to unambiguously determine the source of such differences. The thermal behaviour within the $CaBO_3$ family is rather complex and, consequently, especial attention must be paid when dealing with such compounds as $MgSiO_3$ *ppv* analogues.

References

[1] Murakami M, Hirose K, Kawamura K, Sata N and Ohishi Y 2004 *Science* **304** 855.
Oganov A R and Ono S 2004 *Nature* **430** 445.
[2] Hirai S, Welch M D, Aguado F and Redfern S A T 2009, *Z. Kristallog.* **224** 345.
Hirai S *et al.* 2011 *Chem. Mat.* **23** 119.
[3] Hirose K and Fujita Y 2005 *Geophys. Res. Lett.* **32** L13313.
[4] Niwa K, Yagi T and Ohgushi K 2011 *Phys. Chem. Minerals* **38** 21.
[5] Lindsay-Scott A *et al.* 2011 *J. Appl. Cryst.* **44** 999.
[6] Liu W *et al.* 2011 *Phys. Chem. Minerals* **38** 407.
[7] Lindsay-Scott A, Wood I G and Dobson D P 2007 *PEPI* **162** 140.
[8] Martin C D *et al.* 2007 *American Mineral.* **92** 1048.
[9] Martin C D *et al.* 2007 *American Mineral.* **92** 1912.
[10] Larson A C and Von Dreele R B 2000 *Los Alamos National Laboratory Report LAUR* **86** 748.
[11] Boffa Ballaran T, Trønnes R G and Frost D J 2007 *American Mineral.* **92** 176.
[12] Guignot N *et al.* 2007 *Earth Planet. Sci. Lett.* **256**, 162.
[13] Wang X X *et al.* 2011 *Physica C* **471** 763.
[14] Shirako Y *et al.* 2011 *Phys. Rev. B* **83** 174411.

Journal of Physics: Conference Series **549** (2014) 012025 doi:10.1088/1742-6596/549/1/012025

Cationic superstructures and incommensurate magnetic structure in $SbVO_4$ catalyst

J Hernández-Velasco[1,*], P Vilanova-Martínez[1], J García-García[2], A R Landa-Cánovas[1]

[1]Instituto de Ciencia de Materiales de Madrid (ICMM-CSIC), E-28049 Madrid, Spain.
[2]Centro de Microscopía Electrónica, Univ. Complutense de Madrid (CME-UCM), 28040 Madrid, Spain.

* hernandez.velasco@icmm.csic.es

Abstract. Chemical and magnetic structure modulations in antimony vanadium mixed oxide with composition $Sb_{0.92}V_{1.08}O_4$ prepared in reducing conditions are studied using diffraction techniques, mainly nuclear and magnetic neutron scattering, electron and X-ray powder diffraction. Magnetic susceptibility measurements show possible magnetic order of V^{3+} magnetic moments that is confirmed with neutron diffraction at T_N ~50K. The average commensurate nuclear cell belongs to the tetragonal rutile structural type with cell constants: a=4.6066(11)Å and c=3.0812(8)Å at 60K. However the nature of the magnetic structure is incommensurate with propagation vector $\boldsymbol{k}$= [0, 0, ± 0.266(1)] related to short range order phenomena and V-Sb alternating occupancy along the crystallographic $\boldsymbol{c}$-axis.

1. Introduction

$SbVO_4$ series of compounds could be a promising catalyst for the production of 20% cheaper acrylonitrile by the ammoxidation of propane compared to current method from propylene producing 8×10^6 ton/year, *i.e.* more than 1 Kg per human inhabitant of the planet. In the non-stoichiometric series with ratio Sb:V ~1:1 described as $Sb_{0.9}V_{0.9+x}[]_{0.2-x}O_4$ ($0\leq x\leq0.2$), cation vacancies ([]) are introduced in the basic rutile type-structure following the mechanism: $4V^{3+}\rightarrow 3V^{4+}+[]$, while antimony remains as Sb^{5+}. Composition, oxidation state and subtle structural features influence the activity and selectivity for the highest performance catalyst.

The reduced phase richest in V^{3+} has been reported to be $Sb_{0.9}V_{1.1}O_4$ [1], which shows by electron diffraction superlattice reflections characteristic of a 2-fold rutile superstructure along $\boldsymbol{c}$-axis. However the X-ray powder diffraction pattern showed only basic rutile reflections. The careful reciprocal lattice study by electron diffraction has revealed the following unit cell for the rutile superstructure: $a=\sqrt{2}a_r$, $b=\sqrt{2}b_r$, $c=2c_r$. (subindex r refers to the basic rutile unit cell in $P4_2/mnm$). The superstructure space group, $I4_1md$, was determined by means of convergent beam electron diffraction CBED [1]. A structural model based on alternating Sb and V cations ordering along $\boldsymbol{c}$ in the chains of edge-sharing octahedra was proposed.

No cation vacancies have been observed for this reduced phase, while for the compounds synthesized in oxidizing conditions the presence of vacancies has been confirmed [2]. In fact, electron diffraction experiments have shown that vacancies order in the basic rutile structure giving rise to a modulated structure on the other end member of the series, namely $Sb_{0.9}V_{0.9}O_4$ [3,4]. On the other hand, our magnetic susceptibility data indicate for the first time possible magnetic ordering in reduced phases rich on V^{3+} which are close to the nuclear superstructure by Sb-V ordering. For the simultaneous crystal and magnetic structure determination we have performed a neutron diffraction study on different samples of the series with composition ~$SbVO_4$.

2. Experimental details

The synthesis of the samples in reducing atmosphere has been carried out by mixing stoichiometric amounts of high purity reactants Sb_2O_3 and V_2O_5 in an equimolar ratio (1:1) and heating in a tubular furnace at 800ºC under argon flow with copper filter in two runs of 12 hours with a careful grinding between steps. Samples have been characterized by powder X-ray diffraction in a Powder Diffractometer Bruker D8 Advance with Cu K α radiation.

Electron diffraction experiments were carried out in a JEOL 2000FII transmission electron microscope with a double-tilt holder running at 200kV acceleration voltage.

Bulk magnetic measurements between 1.9K and RT were performed using a SQUID magnetometer MPMS-XL Quantum Design. Isothermal magnetization measurements were collected up to 50 kOe.

Neutron diffraction data between 1.5K and RT were collected on the E6 powder diffractometer at the BENSC nuclear reactor in the Hahn-Meitner-Institut Berlin. The focusing PG-monochromator gave a wavelength λ=2.446(1)Å. Both X-ray and neutron powder diffraction data have been analyzed with the Rietveld method using the program FullProf [5].

3. Results and discussion

3.1. Magnetic measurements

In Figure 1 the magnetic molar susceptibility curve of reduced ~$SbVO_4$ is shown. The net maximum at T=31K could indicate the appearance of three-dimensional antiferromagnetic 3D-AF order within the magnetic V sublattice. The linear fitting of χ^{-1} *vs* T, that follows the Curie-Weiss law, yields a paramagnetic moment value μ_{eff} ~ 2.8μ_B which agrees well with the theoretical moment calculated for the free ion V^{3+} (S=1) following $\mu=2[S(S+1)]^{½}$ = 2.828 μ_B. It is fully compatible with the stoichiometry of the sample $Sb_{0.92}V_{1.08}O_4$ found by simultaneous refinement of X-ray and neutron diffraction data for the atomic occupation factors (Sb/V) and XEDS measurements, which in this case amounts to 0.16V^{4+} (S=½) and 0.92V^{3+} (S=1) per formula unit.

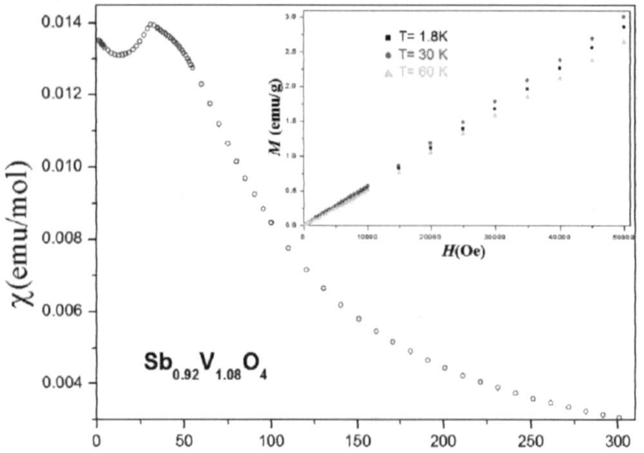

Figure 1. Magnetic susceptibility as function of T measured at an applied field H=500Oe. Inset shows magnetization isotherms at the indicated temperatures.

This bulk properties are in contrast with other ~$SbVO_4$ compositions not prepared under inert atmosphere (Ar or N_2 gas flow) where no magnetic order has been observed. The substitution of V^{3+}(S=1) by V^{4+}(S=½), while the synthesis conditions become more oxidizing, affects the spin arrangement of $Sb_{0.9}V_{0.9+x}[]_{0.2-x}O_4$ in such a way that the magnetic order seems to be destroyed by the appearance of vacancies and the chemical disorder in the cationic positions [6], as we have observed in samples prepared either in air or under oxygen flow.

We collected magnetization isotherms up to H=5T at 10 different temperatures above and below that of the susceptibility maximum. The inset of Figure 1 shows some selected ones, a linear increase of M with magnetic field H is observed in all cases. Anomalies below the Néel temperature are not present like metamagnetic transitions induced by the field neither level-off by saturation.

3.2. Electron and X-ray diffraction measurements

For the reduced compounds ~$SbVO_4$, X-ray powder diffraction patterns might be refined within the rutile structure with random statistical occupancy for Sb/V in *2a* Wyckoff positions (Figure 2). However, electron diffraction patterns show additional periodicities either doubling the unit cell of the basic rutile structure along the crystallographic ***c***-axis (Figure 3 left side), or showing intense diffuse lines running between the rutile-type Bragg maxima (Figure 3 right side).

While alternating -Sb-V-Sb-V- ordered cations along the chains of octahedra sharing edges along ***c*** might explain a commensurate superstructure order, diffuse lines would be compatible maintaining this ordered sequence intrachain but breaking the order between different chains. Furthermore, it has been found some incommensurability due to the imperfect ordering between vanadium and antimony atoms so statistically the periodicity is below $2\times c$ in real space and $\boldsymbol{q} > 1/2\ c^*$ in reciprocal space.

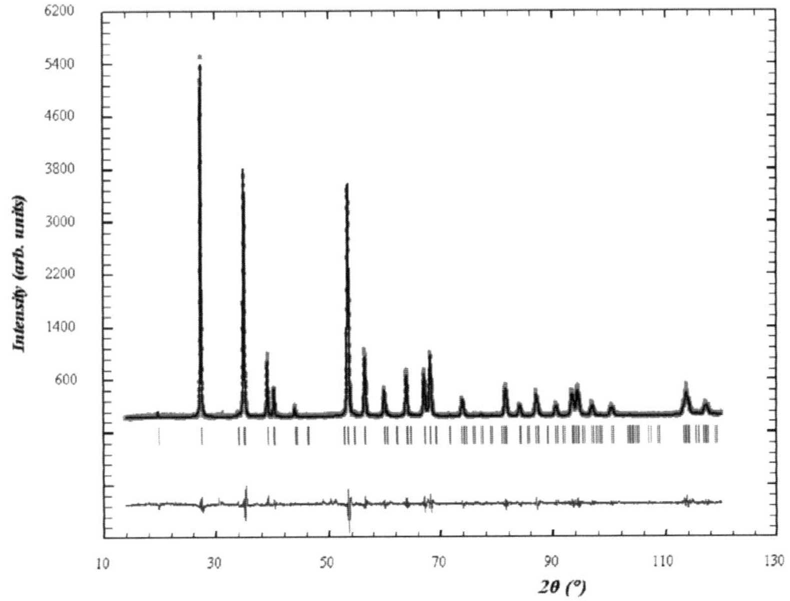

Figure 2. X-ray powder diffraction pattern refined with the Rietveld method for $Sb_{0.92}V_{1.08}O_4$ oxide. Red dots are the experimental data, continuous black line is the calculated profile and blue horizontal line is the difference between them, vertical green ticks mark the allowed Bragg reflections.

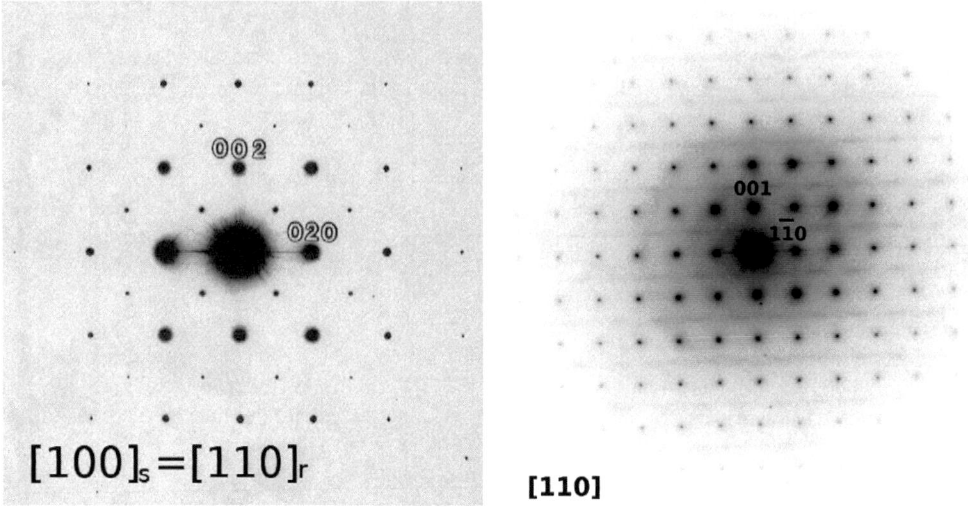

Figure 3. Selected area electron diffraction patterns for reduced compounds ~$SbVO_4$. To the left notice superlattice commensurate reflections in SAED oriented along the [100] zone axis of the supercell equivalent to [110] basic rutile direction. To the right lines of diffuse scattering running perpendicular to the ***c****-axis are observed for the same orientation in another sample.

3.3. Neutron diffraction measurements

Figure 4 shows the neutron diffraction pattern at 1.5K of the sample with composition $Sb_{0.92}V_{1.08}O_4$ where reflections of magnetic origin develop below T_N ~50K [6].

Above the Néel temperature, and taking into account the absence of extra superlattice reflections coming from atomic order, the Rietveld refinement of the pure nuclear structure at T=60K in tetragonal *P4₂/mnm* space group (rutile structure) with position atomic coordinates Sb/V *2a* (0,0,0) and O *4f* (*x*, *x*, 0) *x*=0.3035(5) ; lattice parameters: a_r = 4.6066(11)Å, c_r= 3.0812(8)Å fits the experimental intensities (Agreement factor R_B = 4.2%). The refined fractional cation occupancies are: 0.068(1) and 0.057(1) for V/Sb respectively, normalized using the multiplicity of the general Wyckoff position *i.e. 16k*. The result of simultaneous joint Rietveld refinement of high resolution neutron diffraction and X-ray data at room temperature will be published elsewhere.

Below $T_N \leq$ 50K, new Bragg reflections of pure magnetic nature appear due to the 3D-AF interaction of the vanadium spins changing gradually from the paramagnetic disorder to the magnetically ordered state as the temperature decreases. The periodicity of the magnetic structure develops being close to doubling again the two-fold commensurate supercell of rutile type described by the alternate ordering of Sb and V along the rutile ***c***-axis in $SbVO_4$, giving rise to a final 4-fold rutile superstructure, see schematic representation in Figure 4 inset.

However, the exact periodicity of the spin arrangement is clearly not commensurate with the nuclear cell but close to that. The refinement of the incommensurate magnetic propagation vector to ***k***= [0, 0, $\pm k_z$] with k_z= 0.266(1) *r.l.u.* is close to the commensurate value ¼. It accounts both for the magnetic satellites and it is compatible with the absence of a commensurate nuclear 2-fold supercell. Thus it fully agrees with half the value in reciprocal lattice units and double periodicity in real space to that of the observed modulated chemical structure considering a modulation vector ***q*** = 0.532 *c** measured experimentally in the electron diffraction patterns.

Because the intensity of the magnetic reflections is proportional to the square of the perpendicular component of the magnetic structure factor $\mathbf{F_M}$ to the scattering vector **Q** one might expect that the antiferromagnetic coupling of V^{3+} magnetic moments favours their alignment close to the ***ab*** plane approaching *2S* μ_B value at saturation but with some degree of frustration due to interchain disorder.

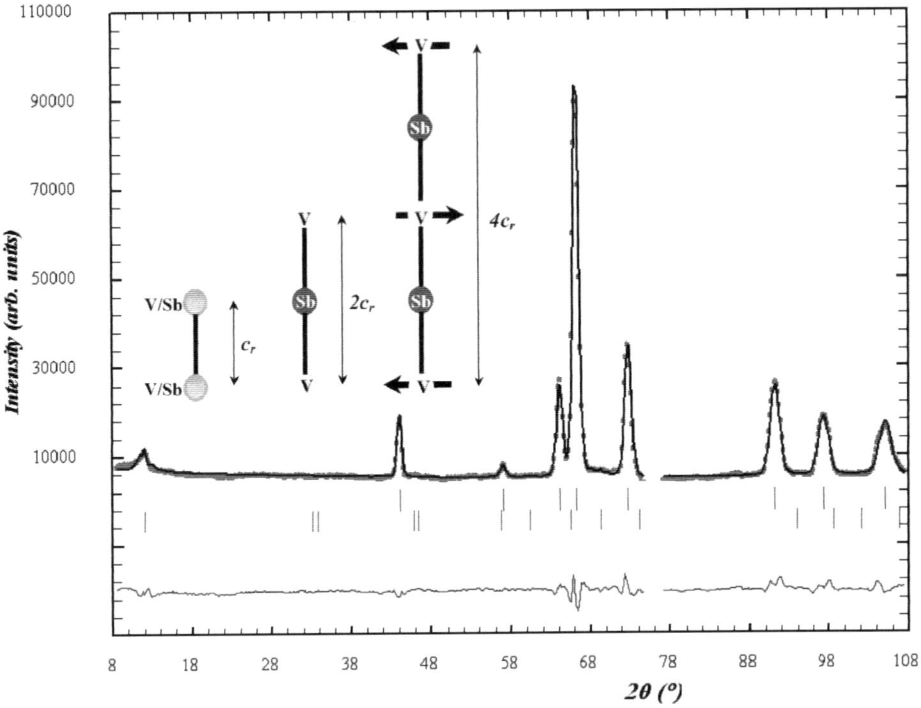

Figure 4. Neutron diffraction pattern of $Sb_{0.92}V_{1.08}O_4$ at T=1.5K refined with the rutile structural model for the nuclear part while the magnetic scattering has been fitted with a profile matching mode. Bottom row of reflections corresponds to the magnetic structure driven by the incommensurate propagation vector described above in the main text.

Nevertheless, the fact that V coherent scattering length for neutrons b_c is rather small give us an advantage to determine the superstructures and incommensurate modulations due to alternating Sb-V occupation of two crystallographic positions split from the substructure in $I4_1md$ and additional information constrained with X-ray data. Also due to the unique property of the neutrons to interact with ordered magnetic moments, the magnetic scattering coming from the V spins could be linked to its crystallographic position, giving more information than at first glance could be expected by its almost null nuclear scattering. In summary, the combination of diffraction techniques gives a better knowledge of the structure to property relationship in this system of high industrial interest.

Acknowledgments
JHV thanks the beamtime provided by HZB, formerly HMI BENSC Berlin through *Eigenforschung* plan and NMI3 support. We thank the MEC funding through projects Consolider CSD2008-00023 FUNCOAT and MAT2011-27192 FLEXOCAT.

References
[1] Landa-Cánovas A R, Hansen S, Ståhl K 1997 *Acta Crystallogr. B* **53**, 221
[2] Hansen S, Ståhl K, Nilsson R, Andersson A 1993 *J. Solid State Chem.* **102**, 340
[3] Landa-Cánovas A R, Nilsson J, Hansen S, Ståhl K and Andersson A 1995 *J. Solid State Chem.* **116**, 369
[4] Landa-Cánovas A, García-García J, Hansen S 2010 *Catal. Today* **158**, 156
[5] Rodríguez-Carvajal J 1993 *Physica B* **192** 55
[6] Hernández-Velasco J, García-García J, Landa-Cánovas A R 2012 *Microsc. Microanal.* **18**, 95

Optimum calcination temperature in the synthesis of a N-C-S co-doped TiO_2 photocatalyst, as monitored by neutron diffraction

Francisco Rozas*, Nicklas Bengtsson*, Oscar Fabelo, Inés Puente**, Marta Castellote***

*Institute of Construction Science Eduardo Torroja (IETcc-CSIC), Serrano Galvache 4, 28033 Madrid (Spain).

** Institut Laue-Langevin BP 156 6, rue Jules Horowitz 38042 Grenoble Cedex 9 France.

Institute of Construction Science Eduardo Torroja (IETcc-CSIC), Madrid, Spain.

f.rozas@ietcc.csic.es

Abstract. Calcination of a co-doped (N-C-S) TiO_2 photocatalyst has been done following the whole process by neutron diffraction (instrument D1B of the ILL) with the objective of study the transition between two crystalline phases: anatase and rutile in order to find the optimum temperature leading to the coexistence of both for this catalyst whose band gap has been shifted to give activity in the visible part of the electromagnetic spectrum. This coexistence is important as it has been established that these catalyst are more efficient than those having only one individual phase. At this respect, it is very important to find the optimum temperature at with synthesis has to be done. In this research it was established that at the heating ramp of 17ºC/5 min, the coexistence of both crystalline phases took place within a narrow margin of temperatures between 610ºC and 690ºC. At higher temperatures only rutile can be identified as the crystalline phase of the TiO_2.

1. Introduction

The discovery of the photosplitting of water in semiconductor-based photoelectrochemical cells by Fujishima and Honda in the earlier 1970 [1] was the origin of a flourishing interest in photocatalysis, initially focused on energy production from water and light and later also on degradation of environmental pollutants in water and air. The crystalline phases of TiO_2 most frequently used in photocatalytic applications are anatase and rutile, with band gaps of 3.26 and 3.05 eV respectively [2], that correspond to absorption of wavelengths in the near-ultraviolet region. This fact cause impediment for the photocatalytic degradation of contaminants under conditions with poor illumination or lack of UV light, for example indoor environments and road tunnels. To enhance the advantage of present and common light sources, the development of visible light active TiO_2 can be of considerable interest.

The red-shift in absorption to visible light of photocatalytic materials can be induced by the incorporation of foreign elements in the semiconductor material. Previously employed methods are impregnation of the surface [3-4], integration of the foreign atoms in substitutional or interstitial positions of the semiconductor crystal structure by either ion implantation [5] or chemical synthesis [4]. This produces changes inside the electronic band structure by formation of intermediate states inside the forbidden gap or by deformation of the band itself. Co-doped titanium active photocatalyst have received much attention during the last decades for its promising impact on environmental protection, and it is known that the microstructure of the photocatalytic materials has a critical role on the photocatalytic activity [6] being the mixture anatase-rutile more active than both phases by itself. The synthesis and the preparation of photocatlytic active titanium dioxide have an important influence on the microstructure but also the calcination procedure plays a significant role, when increasing the temperature the transformation of anatase to rutile is enhanced [7,8]. Therefore it is important to find in each case, the calcination temperature that leads to the formation of the two phases in an optimum way.

In this work, an N-C-S- co-doped photocatalyst synthesized by the authors [9] have been monitored by neutron diffraction under controlled calcination, following the transition between anatase and rutile trying to find the range of coexistence for this catalyst.

2. Experimental

TiO2 based photocatalysts were synthesized following the method described in [10], with some modifications. Titanium isopropoxide was used as precursor and thiourea ($(NH2)_2CS$) was used to introduce dopants into the crystal network of TiO_2, provided that it contains the three species.
The synthesis was as follows: pure thiourea (53.6 g) was dissolved during continuous stirring in 700 ml ethanol, heated to 65ºC. Titanium isopropoxide (50 g) was then added slowly and let to stir during 1 h. The solution was concentrated under reduced pressure resulting in the formation of white slurry that was left to dry in room temperature for 2 days to obtain a white powder.

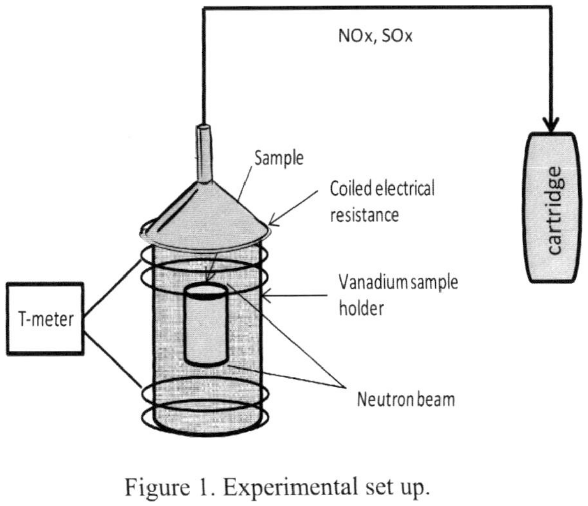

Figure 1. Experimental set up.

Samples were calcinated at the CRG-two-axis diffractometer D1B place in the Institut Laue-Langevin in Grenoble (France), being monitored on site by taking continuous diffraction patterns of the process. A schematic representation of the experimental device can be seen in Fig. 1.

A coiled electrical resistance, connected to a temperature programmer and to a data logger was used to heat the sample. It was opened resulting in a gap in its structure in order to avoid overlapping of their peaks in the spectrum acquisition. During the experiment they were kept inside of a vanadium sample holder with a high of 10 cm, an external diameter of 8.5 cm and an internal diameter of 4.8 cm. It was also tested a sample holder made of ceramic; however, it was discarded because the neutron diffraction spectrum of the catalyst was highly influenced by the composition of the ceramic. The sample was connected to a cartridge of activated charcoal in order to capture the SOx and NOx gasses being resulting from the calcinations. This is the reason why a conventional furnace was not used, to be able to safely recover the gasses. Heating followed a temperature ramp: 17ºC/5 min, until 820°C.

D1B diffractometer was chosen because of its high efficiency position sensitive detector (PSD) which makes it suitable for real time experiments and for very small samples. The vertical distribution of neutron beam during the experiment was wide enough to cover the whole sample. The wavelength of the experiment was 2,52Å.

3. Results and discussion

Several peaks were identified during the different steps of the experiment; some of them were due to the presence of the vanadium sample holder, others correspond to the different crystalline phases of the titanium dioxide. In Figure 2 an example of diffraction patterns at different temperatures in the region between values of 2 theta of 60 and 74 (in order to avoid overlapping) are presented. As can be seen in Fig. 2 at the beginning of the experiment (T=25°C) there are no peaks in the region of interest. During the experiment the shape of the diffraction patterns varies and a small peak can be seen growing when the temperature reaches 450°C at 2 theta value between 66 and 67. This peak can be an indicative of the formation of anatase.

The transition between both crystalline phases con be established when the temperature reaches 653°C. In the pattern epicted in Fig. 2 it can be seen a remaining peak corresponding to anatase and two new peaks corresponding to the formation of rutile (2 theta = 67.2 and 71). At the end of the experiment (T = 820°C) there is no anatase in the sample and another peak corresponding to rutile appears (2 theta = 69.3). It is also represented the evolution of the diffraction pattern when cooling noting that the most intense rutile peaks remain without presence of anatase.

According to the results obtained it can be said that there is a coexistence of both crystalline phases in a very small range of temperature, between 610°C and 690°C. This coexistence disappears when heating.

Figure 3 represents the projection in two dimensions of the experiment since the beginning till the end of the experiment in function of the temperature in °C where, marked as a shaded band, it can be seen the narrow margin of coexistence of both crystalline phases, between 610°C and 690°C.

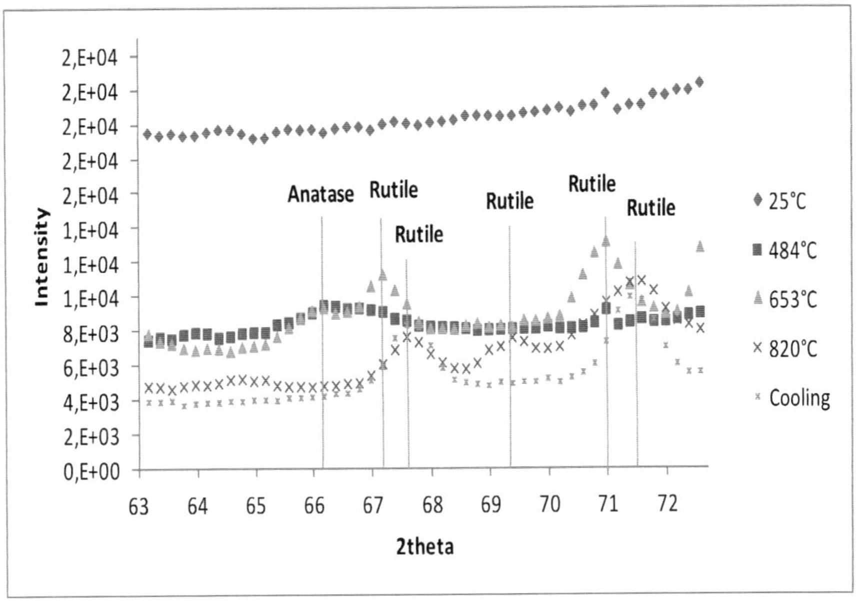

Figure 2. Diffraction patterns at different temperatures.

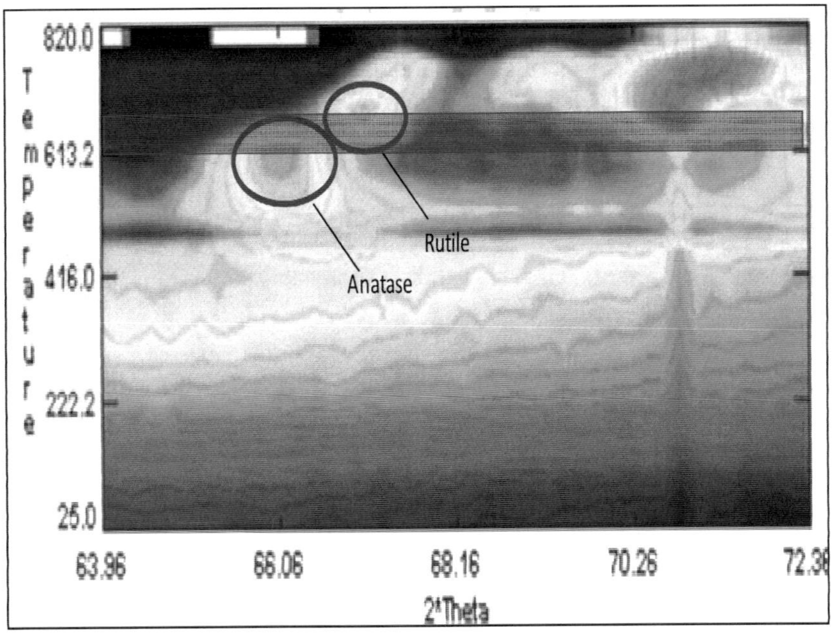

Figure 3. Contour plot of the experiment. The range of coexistence of anatase and rutile has been marked as a shaded band.

4. Conclusions

In this research the evolution of the phase transition anatase → rutile in a co-doped N-S-C titanium active photocatalyst was followed by means of neutron diffraction. At the heating ramp of 17ºC/5 min, the coexistence of both crystalline phases was established within the margin of temperatures between 610ºC and 690ºC, disappearing when heating and remaining only the crystalline phase rutile.

Acknowledgements

The authors are acknowledged for the provision of neutron beam time to the D1B of ILL.

References

[1] Fujishima, A.; Honda, K. *Nature 1972*, 238, 37-38.
[2] Carp, O.; Huisman, C. L.; Reller, A. *Progress in Solid State Chemistry* 2004 32 33-177
[3] Anpo, M.; Takeuchi, M. *International journal of Photoenergy* 2001, 3, 89.
[4] Di Valentin, C. ; Finazzi, E.; Pacchioni, G.; Selloni, A.; Livraghi, S.; Paganini, M. C. et al. *Chemical Physics* 2007, 339, 44-56.
[5] Anpo, M.; Takeuchi, M.; Ikeue, K.; Dohshi, S. *Current Opinion in Solid State and Materials Science* 2002, 6, 381-388.
[6] Madhusudan Reddy K Gopal Reddy C Vand Manorama S V 2001 Preparation, Characterization and Spectral Studies on Nanocrystalline Anatase TiO_2. *Journal of Solid State Chemistry*, Vol. 158, pp. 180-186.
[7] Ovenstone J and Yanagizawa K 1999 Effect of Hydrothermal Treatment of Amorphous Titania on the Phase Change from Anatase to Rutile during Calcination *Chem. Mater.* 11, 2770.
[8] Yanagizawa K Yamamoto Y Feng Q and Yamasaki N 1998 Formation mechanism of fine anatase crystals from amorphous titania under hydrothermal conditions *J. Mater. Res.* 13 825.
[9] Bengtsson N Castellote M López-Muñoz MJ Cerro L 2009 Preparation of co-doped TiO_2 for photocatalytic degradation of NO_x in air under visible light. *J. Oxid. Technol.* Vol. 12 No. 1
[10] Ohno T Akiyoshi M Umebayashi T Asai K Mitsui T and Matsumura M. 2004 Preparation of S-doped TiO_2 photocatalysts and their photocatalytic activities under visible light *Applied Catalysis A*: General Vol. 265, 115-121.

Journal of Physics: Conference Series **549** (2014) 012027 doi:10.1088/1742-6596/549/1/012027

Residual strain scanning of alumina-based ceramic composites by neutron diffraction

J. Ruiz-Hervias[1], G. Bruno[2], S. Bueno[3], J. Gurauskis[3], C. Baudín[3], K.Y. Fan[1]*

[1]*Materials Science Department, Universidad Politécnica de Madrid, Spain*

[2]*Corning Incorporated, Corning, NY, 14831, USA*

[3]*Instituto de Cerámica y Vidrio (CSIC), Madrid*

*E-mail: kunyang.fan@mater.upm.es

Abstract. Residual strain profiles were measured by neutron diffraction in alumina-aluminum titanate ceramic composites sintered at two different temperatures, namely 1450 and 1550°C. The results show that irrespective of the direction and the sintering temperature, the obtained profiles are almost flat, with very similar results for both temperatures. In addition, the results demonstrate that the alumina is in compression whereas the aluminium titanate is subjected to tensile residual stresses.

1. Introduction

Ceramic composites have been investigated during decades to obtain improved mechanical performance. Some ceramic composites (Alumina Al_2O_3 - aluminium titanate Al_2TiO_5 composites) offer improved flaw tolerance and toughness [1-4], which is thought to be due to thermally-induced residual stresses. The average crystallographic thermal expansion of aluminium titanate is slightly higher than that of alumina ($\alpha_{A25\text{-}1000°C} = 8.7 \cdot 10^{-6}$ °C^{-1}, $\alpha_{AT25\text{-}1000°C} = 9.7 \cdot 10^{-6}$ °C^{-1}) and it is highly anisotropic. As a consequence, in an alumina-aluminium titanate composite, high tensile and/or compressive residual stresses are expected to develop among grains, depending on the particular relative orientations. These stresses might lead to a wide variety of possible toughening mechanisms: crack deflection, crack branching, crack bridging and microcracking. Therefore, the characteristics of the toughness curve and, as a consequence, the flaw tolerance of alumina-aluminium titanate composites may be strongly modified by appropriately altering the microstructure: grain size and aluminium titanate content.

Residual stresses have been measured by neutron diffraction in several ceramic systems [5-7]. However, residual stress profiles are seldom reported [7].

In this work, residual strain profiles were measured by neutron diffraction in 60 vol.% Al_2O_3+40 vol.% Al_2TiO_5 ceramic composites (from now on A-AT40) prepared by slip casting and sintered at two different temperatures, namely 1450 °C and 1550 °C.

2. Experimental

2.1. Material

The investigated Al_2O_3/Al_2TiO_5 composites with 40 vol. % of Al_2TiO_5 were obtained using α- Al_2O_3 (Condea, HPA05, Houston, USA) and previously reacted Al_2TiO_5 powders fabricated using the same alumina and high purity anatase-TiO_2 (Merck, 808, Germany) powders. Plates with 70mm×70mm×10mm dimensions were obtained by slip casting, removed from the moulds and dried in air at room temperature for at least 24 h. In order to obtain the monolithic materials, the dried blocks were sintered in an electrical furnace (Termiber, Spain). The samples were heated at a rate of 2 °C/min, with 4 hours dwell at 1200 °C during heating and 3 hours dwell at the sintering temperature. Two different sintering temperatures, namely 1450 °C and 1550 °C, were used in order to obtain two different sets of samples. Sintered pieces were then machined to obtain test samples with final geometry of 30mm×30mm×10mm.

2.2. Residual strain measurements

Residual strain measurements were performed by neutron diffraction on the strain imager SALSA (Strain Analyser for Large Scale engineering Applications) [8], at the ILL, Grenoble, France. SALSA features a Stewart platform as a sample stage, which allows very flexible sample handling capabilities and arbitrary scan trajectories in the 3D space. In our case, the (311) reflection of the double-focusing Si-monochromator was used at a take-off angle of 85° to obtain a wavelength of λ = 2.06 Å. A 2D position-sensitive detector with an angular opening of approximately 5° was mounted on SALSA. Primary and secondary slits were used to set the gauge volume (which can be varied automatically via step motors) of 1.5mm×1.5mm×10mm. Strain scanning was carried out in both phases, Al_2O_3 and Al_2TiO_5, in parallel (parallel to the largest surface of samples) and normal directions (perpendicular to the largest surface of the samples), with a scan step of 1 mm, as shown in Fig. 1.

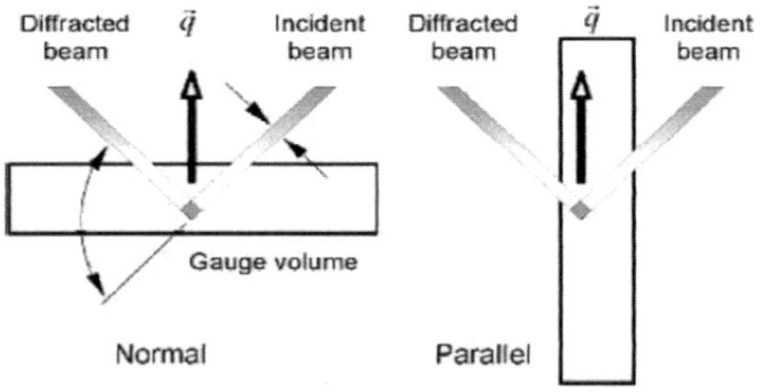

Fig. 1. Experimental setup employed in the neutron diffraction experiment for the normal and parallel directions

The (116) reflection was selected for Al_2O_3, which corresponds to 2θ=80.0° for the above-mentioned wavelength. For Al_2TiO_5, the (153) peak at 2θ=87.9° was selected, which is accompanied by a neighbouring peak (062) at 2θ=87°. The LAMP software [9] was used to analyse the experimental results and obtain the peak parameters at each sample orientation.

According to Bragg's law, by precise measurements of the diffraction peak shift, the variation in the inter-planar spacing for a given {*hkl*} set of planes can be evaluated and thus the elastic strain within the material can be calculated:

$$\varepsilon_{hkl} = \frac{d_{hkl} - d_{0,hkl}}{d_{0,hkl}} \tag{1}$$

where ε_{hkl} is the longitudinal strain in the direction of the scattering vector, and $d_{0,hkl}$ is the unstressed lattice spacing of the *hkl* reflection. The later was determined by measuring the single phase alumina and aluminium titanate powders employed to fabricate the samples and the results are reported in Table 1.

Table 1. Unstressed lattice spacing values of the *hkl* reflections employed

	$Al_2O_3(116)$	$Al_2TiO_5(062)$	$Al_2TiO_5(153)$
$d_{0,hkl}$ (Å)	1.60365±0.00014	1.4959±0.0006	1.4840±0.0003

3. Results and discussion

The residual strain profiles along the sample thickness are depicted in Fig. 2, for the two sintering temperatures, namely 1450ºC (Fig. 2a) and 1550ºC (Fig. 2b). The strain results for both sintering temperatures are very similar. This suggests that an increase of sintering temperature from 1450ºC to 1550ºC has no clear effect on residual strains for alumina-aluminium titanate monolithic composites.

Linear elasticity was employed to calculate residual stresses from residual strains [10]. The diffraction elastic constants of the alumina phase were calculated by a Kröner average for the (116) reflection ($E_{A,116}$ = 407 GPa, $\nu_{A,116}$ = 0.21), from the single crystal data published in the literature [11]. Since there are no reliable plane-specific diffraction elastic constants available for Al_2TiO_5, bulk elastic constants (E_{AT} = 155 GPa, ν_{AT} = 0.33), were used in the present work [12]. For the residual stress calculation, the (153) reflection of Al_2TiO_5 was chosen, because it was more intense and narrower than the (062) reflection.

Residual stresses profiles for the sample sintered at 1450ºC are shown in Fig. 3. Stresses are not reported for the sample sintered at 1550ºC because there are not enough measurements in the parallel direction. It can be seen that the stress profiles are almost flat, with tensile residual stresses in Al_2TiO_5 particles (average around 500 MPa) and compressive ones in the Al_2O_3 matrix (average around -160 MPa). The differences between normal and parallel stresses lie within experimental error in all cases. This suggests a hydrostatic residual stress state in both phases, which is consistent with the fabrication process.

Consequently, the calculated phase stresses should verify the equilibrium condition, i.e. negligible macro residual stresses [10]:

$$\sigma_M = f_A \sigma_A + f_{AT} \sigma_{AT} = 0 \tag{2}$$

where σ_M is the macro residual stress, f_A and f_T are the volume fractions of Al_2O_3 and Al_2TiO_5, respectively, and σ_A and σ_{AT} are the corresponding phase stresses. From Fig. 3 it can be concluded that the calculated residual stresses approximately verify the equilibrium condition (Eq. 2).

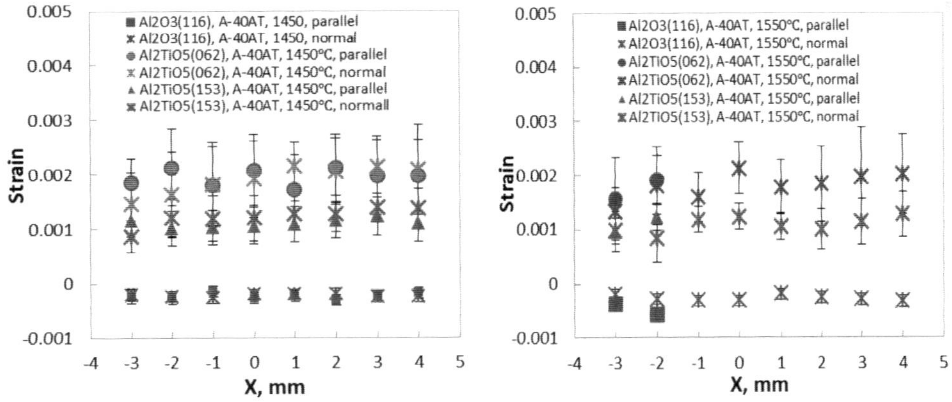

(a) (b)

Figure 2. Residual strain (parallel and normal directions), (a) 1450 °C and (b) 1550 °C.

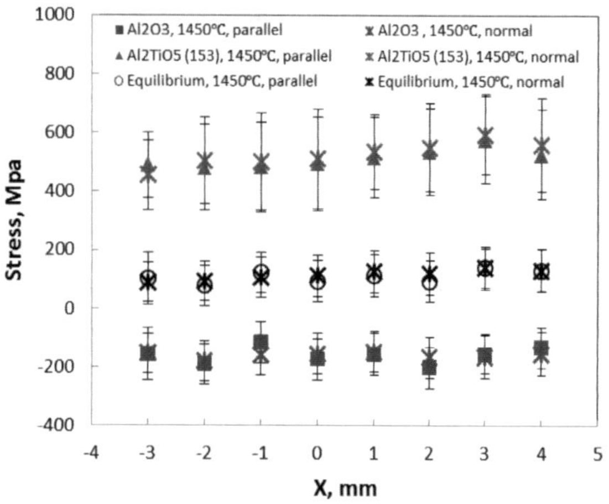

Figure 3. Residual stresses in parallel and normal directions for composites sintered at 1450 °C.

4. Conclusions

Residual strain scanning using neutron diffraction was used to calculate the residual stress profile in the alumina and aluminium titanate phases of Al_2O_3+40 vol. % Al_2TiO_5 ceramic composites. The results show that alumina is under compression (around -160 MPa) and aluminum titanate is under tension (around 500 MPa). The residual stress state is hydrostatic in both phases. A change in the sintering temperature from 1450°C to 1550°C does not have a remarkable influence on residual strains. The calculated macro residual stresses are negligible for the sample sintered at 1450°C.

5. Acknowledgements

This work was supported by the project BIA2011-26486. The authors wish to thank the EPSRC and the ILL for beamtime in SALSA.

References

[1] Runyan JL, Bennison SJ 1991 *J. Eur. Ceram. Soc.* **7** 93.
[2] Padture NP, Bennison SJ, Chan HM 1993 *J. Am. Ceram. Soc.* **76** 2312.
[3] Bartolome J, Requena J, Moya JS, Li M, Guiu F 1996 *Acta Mater.* **44** 1361.
[4] Bueno S, Moreno R, Baudín C 2004 *J. Eur. Ceram. Soc.* **24** 2785.
[5] Wang XL et al. 1994 *J. Am. Ceram. Soc.* **77** 1569.
[6] Choo H, Bourke M, Nash P, Daymond M, Shi N 1999 *Mats. Sci. Eng. A* **264** 108.
[7] Lukas P, Vrana M, Mikula P, Vleugels J 2004 *Physica B* **350** e517.
[8] Pirling T, Bruno G, Withers PJ 2006 *J. Mats. Sci. Eng. A* **437** 139.
[9] Richard D, Ferrand M, Kearley GJ 1996 *J. Neutron Research* **4** 33.
[10] Hauk V 1997 *Structural and residual stress analysis by nondestructive methods* Elsevier.
[11] Gnäupel-Herold T, Brand PC, Prask HJ 1998 *J. Appl. Cryst.* **31** 929.
[12] Bueno S, Hernández MG, Sánchez T 2008 *J. Eur. Ceram. Soc.* **34** 181.

Neutron diffraction as a tool in the study of reinforced concrete. Compilation of some cases

M. Castellote

Institute of Construction Science Eduardo Torroja (CSIC). Serrano Galvache 4, 28033 -Madrid- Spain

martaca@ietcc.csic.es

Abstract. Cementitious materials are much more complex than it seems at a first sight. On one hand, due to the excess of water needed to make the mix workable, a network of pores is generated, that puts in contact the material with the environment and allows their attack by aggressive agents that can be physical of chemical agents, producing the deterioration of the concrete itself and corrosion of the rebars. Then, it is necessary to study the transport properties and the chemical and physical interaction of aggressive agents with the solid and liquid phases of the cement paste the corrosion of the rebars, and the reparation processes. This is an approach concerning the service life of structures, however, we cannot forget an important chapter which is gaining much relevance in the last time: that of the special uses of concrete, for which, tailored concretes have to be designed. In order to undertake these problems, we need new analysis tools, different that the traditionally ones applied to study concrete, that allows the understanding of the mechanisms regulating the processes. One of these analysis tools is neutron diffraction that gives us the possibility of study the bulk of materials using a quite big specimen. In this work, 4 different problems undertaken with the help of neutrons in experiments carried out by the group of the author at the ILL, at the D1B and D20 instruments are presented.

1. Carbonation studies

Carbonation can be defined as the neutralisation of the cement paste due to reaction with acid atmospheric agents, mainly CO_2. When the neutral front arrives to the rebar in reinforced concrete, depasivation of the steel takes place and corrosion of the steel starts. Deepen in the knowledge of this process was undertook by in-situ monitoring of the changes that take place in the phase composition of cement pastes during accelerated carbonation (100% CO_2) at 65% HR for different binders (plain OPC and OPC with additions of fly ash and micro-silica) by taking Neutron Diffraction patterns in parallel with the carbonation experiments carried out at the D20 (a schema of the experiment is given in figure 1-a) [1]. The variation of the intensity of chosen reflections for each phase along the experiment was used to monitor concentration changes supplying data, in real time, for establishing the fractional conversion of different phases (Portlandite, Ettringite and CSH gel) of the hydrated cement pastes. Fitting of these results allowed to make a qualitative approach to the kinetics of the carbonation of the different phases deducing that all the phases consumed by the carbonation process disappear according to an exponential decay of first order, while formation of Calcite follows the trend of a sigmoid shape given by a Boltzmann function, as corresponds to a finite amount of reacting material (figure 1-b).

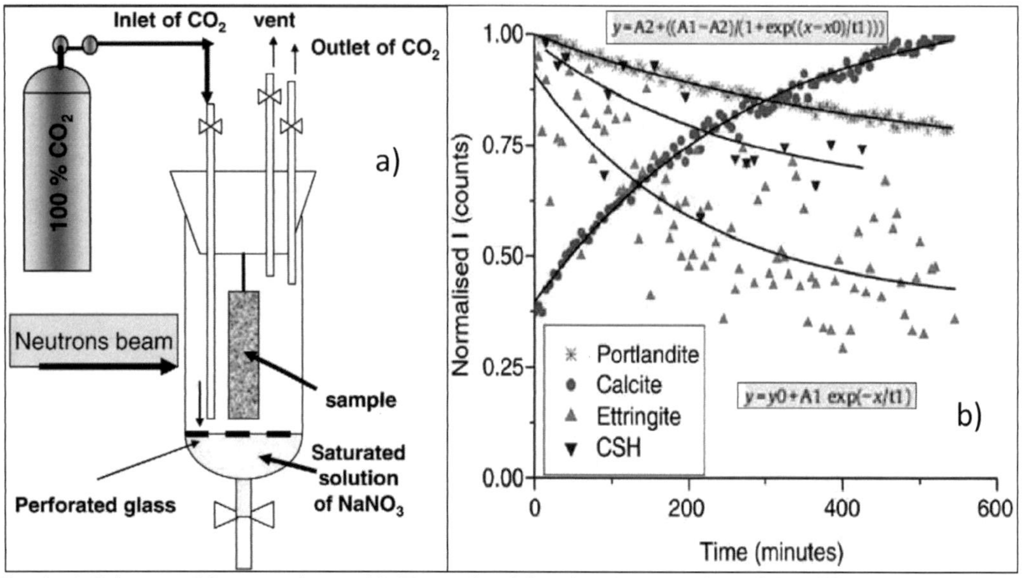

Fig. 1: a) Schema of the experiment. b) Example of fractional conversion of the different phases for the mix of OPC and fitting to the equations.

It could be deduced that the rate of decrease of each phase is different for the different mixes. For the same mix, the rate for the different phases is different, but maintaining the same proportionality between each phase in the different samples. Ettringite is the phase that reacts quicker, then, the crystalline fraction analysed of the CSH gel, being the consumption of Portlandite, the slowest persistent process. Additionally, interesting conclusions on the microstructural changes that takes place during the carbonation of cement pastes were extracted. Details on this experiment, treatment of the data and conclusions can be found in [1]. Additionally, the evolution of cement paste samples during accelerated carbonation was followed in situ, at the D1B instrument, at different RH, being able to find an optimum of the 53% of RH. In combination with other techniques, the presence of initial amorphous calcite was evidenced, as well as the crystalline character of the new calcite formed after carbonation [2].

2. Electro-osmosis and alkaline reserve restoration during realkalisation

Realkalisation is an electrochemical treatment for restoring to concrete, which surrounds reinforcing bars, a high pH value corresponding to sound-non carbonated concrete [3]. Even though the electroosmotic flux through hardened cementitious materials during laboratory realkalisation trials had been previously noticed, there was some controversy on either it was really an electrokinetic phenomena or just salvation water of Na^+ ions. In-situ monitoring of the experiment, analysing at the same time the establishment of the electroosmotic flux and the microstructure changes in the surroundings of the rebar was possible through neutron diffraction experiments: two series of cement pastes, cast with CEM I and CEM I substituted in a 35% by fly ash, previously carbonated at 100% CO_2, were submitted to realkalisation treatments followed on line by simultaneous acquisition of neutron diffraction data [4]. A ponding was glued on the top of the cylinder of cement paste (prepared with deuterated water) and was filled with a solution of Na_2CO_3 1M prepared with ordinary, non-deuterated, distilled water. The experiment was designed on the hypothesis that if electroosmotic flux took place, an abrupt increase in the background of the diffraction patterns should occur due to incoherent scattering of the hydrogen of the plain water.

In order to determine the exact moment, and therefore the conditions in the experiment in which the electroosmotic flux starts to develop, the derivative of the curve of the background in function of time has been calculated for both experiments. Therefore, in the inflexion points of the original curve towards higher slopes, a maximum appears in the derivative curve. The positions of these maximums give the points of establishment of the electroosmotic flux.

The averaged values of zeta-potential, calculated from the averaged electroosmotic flux, for mixes A and B were also calculated, being of -1.2 and -4 mV, respectively. Therefore, the zeta potential of the electrical double layer in the walls of the pores of cement paste is more negative when adding fly ashes to the clinker than when using plain CEM I. Concerning the evolution of phases, it could be directly observed the precipitation of portlandite on the rebar and on the zone close to the negative electrode as far as the pH was high enough. The provision of Ca^{2+} ions come from the dissolution of calcite and vaterite. So, for the first time, it could be said that it is possible to restore the alkaline reserve if passing enough amount of charge during a realkalisation treatment.

3. Resistance in case of fire

Other important aspect is the resistance of the structures in case of fire. In this framework, composition and microstructure changes of cement pastes when heating until 620 ºC and cooling afterwards were monitored on site by neutron diffraction at the D1B (ILL). The parameters involved in the study were the heating ramp, the state of the sample (in block or ground) and the type of cement. The residual state of the samples was complemented by mercury intrusion porosimetry (MIP) and thermogravimetry analysis (TGA) [5]. The use of neutron diffraction allowed monitoring the major features of the experiments, i.e., the phases existence domains and their growing and decaying: portlandite, Ettringite, CSH, calcite and larnite. Additionally, it could be stated that Portlandite that was formed when cooling is less crystalline than the original one, and its temperature of thermal decomposition gets lower. On the contrary, calcite that precipitates decomposed thermally at higher temperature than the original one. The monitoring of the loss of free and combined water, by integration of diffraction patterns backgrounds as a function of the internal temperature of the samples could be related to the resulting microstructure. Finally, changing the heating rate, from 60 to 120 ºC/h had no significant influence neither in the composition nor in the resulting microstructure of the cement paste; however, the type of cement or the state of the sample (powder or block) was a significant parameter, leading to different residual phases (free lime content), which might have a significant influence in the durability of a concrete after a fire.

4. Hydrogen embrittlement of high strength steel

Embrittlement of high-strength steel is known to be caused by hydrogen penetration into the network of the metal that could lead to disastrous failures of pre and post tensioned structures. However, the penetration mechanism of hydrogen into the steel is still under discussion, as it is not so easy to detect and measure the hydrogen in steels. To undertake this subject, relatively large commercial cold drawn pearlitic steel wires were submitted to stress corrosion cracking under different environments: inert (air) environment; at a controlled anodic potential of -300 mV (SCE) in 0.05 M aqueous solution of $NaHCO_3$ and at a controlled cathodic potential of -1200 mV (SCE) in 0.05 M aqueous solution of $NaHCO_3$. From the slow strain rate testing (SSRT) of the steels it was shown that both samples in aggressive environment, exhibited brittle fracture, while the inert sample did not. The steel tested at -300 mV (SCE) was the one that presented a higher embrittlement degree. Neutron Diffraction Technique was used to analyze the samples. The data acquisition was performed at the D20 instrument of the (ILL), in Grenoble, France [6]. The wavelength used was 1.3 A, exploring an angular domain in 2θ from 10° to 150°. Cell parameter variations of the different phases were measured using pseudo–Voigt functions and by Rietveld refinement using the Fullprof software. The penetration of the hydrogen into the unit cell of the ferrite was deduced from the shift of the d-spacing after the experiments. The increase of volume in the ferrite unit cell was quantified and correlated with the degree of embrittlement of the steels in aggressive environment (see figure 2).

Other examples of application of neutron techniques to the study of concrete can be found in [7-14].

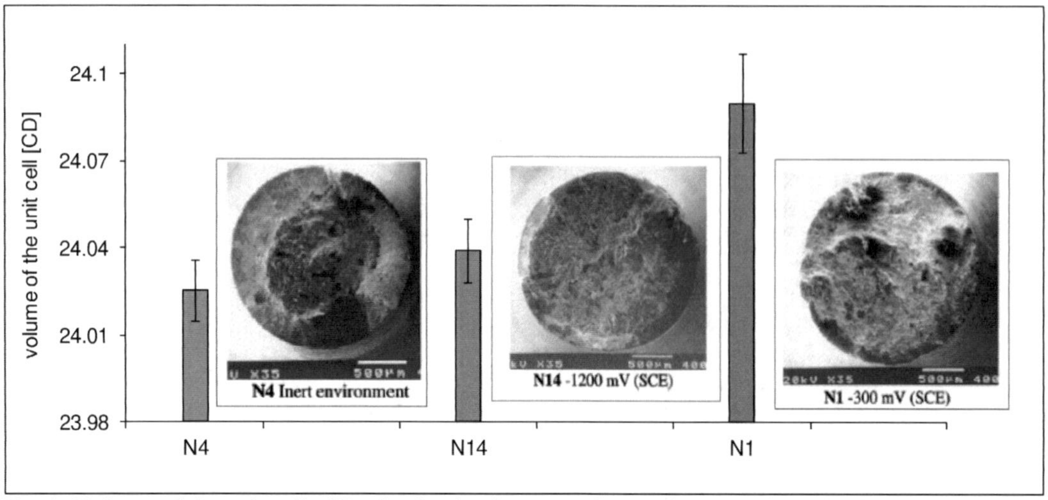

Fig. 2. Volume of the unit cell of Ferrite for the three samples studied, calculated by fitting isolated peaks to pseudo–Voigt functions, and micrograph of the surface of fracture after the tests.

Acknowledgements:
The author is grateful to the ILL, especially to the staff of D1B and D20.

References

[1] M. Castellote, C. Andrade, X. Turrillas, J. Campo, G.J. Cuello, Accelerated carbonation of cement pastes in situ monitored by neutron diffraction. Cem. Concr. Res. 38 (2008) 1365-1373.

[2] Galan, C. Andrade, M. Castellote, N. Rebolledo, J. Sanchez, L. Toro, I. Puente, J. Campo, O. Fabelo, Neutron diffraction for studying the influence of the relative humidity on the carbonation process of cement pastes. J. Phys.: Conf. Ser. 325 (2011) 012015.

[3] CEN/TS 14038 "Electrochemical realkalization and chloride extraction treatments for reinforced concrete" Part-1: Realkalization (2004) .

[4] M. Castellote, I. Llorente a, C. Andrade, X. Turrillas, C. Alonso, J. Campo, In-situ monitoring the realkalisation process by neutron diffraction: Electroosmotic flux and portlandite formation, Cement and Concrete Research 36 (2006) 791 – 800.

[5] M. Castellote, C. Alonso, C. Andrade, X. Turrillas, J. Campo, Composition and microstructural changes of cement pastes upon heating, as studied by neutron diffraction, Cement and Concrete Research 34 (2004) 1633–1644.

[6] M. Castellote, J. Fullea, P.G. de Viedma, C. Andrade, C. Alonso, I. Llorente, X. Turrillas, J. Campo, J.S. Schweitzer, T. Spillane, R.A. Livingston, C. Rolfs, H.-W. Becker, Hydrogen embrittlement of high-strength steel submitted to slow strain rate testing studied by nuclear resonance reaction analysis and neutron diffraction, Nuclear Instruments and Methods in Physics Research B 259 (2007) 975–983.

[7] F. Häußler, F. Eichhorn, S. Röhling, H. Baumbach, Monitoring of the hydration process of hardening cement pastes by small-angle neutron scattering, Cement and Concrete Research, Volume 20, Issue 4, July 1990, Pages 644-654.

[8] R. Berliner, M. Popovici, K. Herwig, H.M. Jennings, J. Thomas, Neutron scattering studies of hydrating cement pastes, Physica B: Condensed Matter, Volumes 241–243, December 1997, Pages 1237-1239

[9] De La Torre, S. Bruque, J. Campo, M.A.G Aranda, The superstructure of C3S from synchrotron and neutron powder diffraction and its role in quantitative phase analyses, Cement and Concrete Research, Volume 32, Issue 9, September 2002, Pages 1347-1356.

[10] M. Castellote, C. Alonso, C. Andrade, J. Campo and X. Turrillas; In-situ hydration of Portland cement monitored by neutron diffraction. Applied Physics A, 74 [Suppl.] S1224-S1226, (2002).

[11] M. Castellote, X. Turrillas, I. Llorente, C. Alonso, C. Andrade and J. Campo; A neutron diffraction study of changes induced in aluminous cement paste by the application of external electrical fields" Physica B 350 (2004) e561-e564

[12] Phillip C. Womble, Jon Paschal, Ryan Moore, Cement analysis using d + D neutrons, Nuclear Instruments and Methods in Physics Research Section B: Beam Interactions with Materials and Atoms, Volume 241, Issues 1–4, December 2005, Pages 765-769

[13] Vanessa K. Peterson, Dan A. Neumann, R.A. Livingston, Hydration of cement: The application of quasielastic and inelastic neutron scattering, Physica B: Condensed Matter, Volumes 385–386, Part 1, 15 November 2006, Pages 481-486

[14] Andrew J. Allen, Jeffrey J. Thomas, Analysis of C–S–H gel and cement paste by small-angle neutron scattering, Cement and Concrete Research, Volume 37, Issue 3, March 2007, Pages 319-324

AUTHOR INDEX

AUTHOR INDEX

Institute of Physics
Dirac House, Temple Back
Bristol BS1 6BE UK

CURRAN ASSOCIATES INC
proceedings.com